Photovoltaics

For The Technically Ungifted,

Academics With Two Left Hands

And Everybody Else

Frontcover:

>   http://commons.wikimedia.org/wiki/File:Energia_renovable.jpg
>   Title: „Paneles Solares en Guantanamo"
>   The photo was set public domain by Carla
>   (I could not find out more about the photographer)

Translation to English:

>   Jürgen D. Henning
>   Jan Woudhuysen was so nice (and extremely patient) to convert most my text into proper English.

This work is under copyright law. Electronic copies only with previous permission. Photo mechanic copies for teaching purposes will be tolerated. All pictures and graphics were either done by me or they are labeled and used with permission.

Bibliographic Information of the German National Library:

The German National Library enlists this publication in the German National Bibliography; detailed information can be obtained via the Internet from www.dnb.de

© 2014 Jürgen D. Henning
Production and Publishing:
BoD – Books on Demand, Norderstedt

ISBN: 9783735758903

Table of Content

Foreword 7

1. Should this book be available in your own language? 11

2. Quick look at the German 'Energiewende' 13

3. The costs of electricity 17
   3.1 The electricity price in Germany 18
   3.2 The situation today (end of 2013) 22

4. A minimum of physics 23

5. Costs running a generator 28

6. Its possible: one kWh for 2.6 Euro cents! 32

7. Drunk, stupid - or simply criminal? 35
   7.1 First of all: what's going on? 35
   7.2 My home is my castle! 37
   7.3 Desertec 39

8. Photovoltaics and the seasons 42
   8.1 A few examples of irradiation 44
   8.2 The 37°-belt 46
   8.3 Enlarged zone: the 55°-belt 47
   8.4 Case studies: two examples from the 55° belt 49

9. Are PV installations dangerous? 56
   9.1 How could a flashing arc start? 57
   9.2 How to switch off segments safely 59
   9.3 Cable cross sections 59
   9.4 Energy losses in the cables 60
   9.5 Connectors: MC3 & MC4 and Sunclix 67
   9.6 Fuses 70
   9.7 Safety regulations 71

10. You want to give it a try? 73

    10.1 What did we gain?..78
    10.2 The Paraffin Lamp..79
    10.3 Intermediate results...82

11. The solar cell..84
    11.1 About light and shadow..85
    11.2 Shorted solar panels. Dangerous?...89
    11.3 Wp and the other cryptic information.......................................89
    11.4 How to find the right panels?...92
    11.5 Tracking-Systems..94

12. Solar controller..96
    12.1 Direct connection..98
    12.2 Shunt regulator..100
    12.3 The linear regulator...101
    12.4 PWM regulator..102
    12.5 Energetic differences: linear and PWM regulators....................103
    12.6 The MPPT regulator..105
    12.7 Comparison: PWM regulator vs. MPP regulator.......................109
    12.8 Addendum: direct connection / shunt PWM............................111

13. Modifying solar installations..112
    13.1 Aging of panels..114
    13.2 Connect voltage sources in parallel...115
    13.3 Voltage source interconnected in line.......................................116
    13.4 Reducing line losses...117
    13.5 Charge controllers and higher voltages....................................119
    13.6 Light and shade again...120
    13.7 Several charge controllers to one battery?................................121

14. Storing electric energy...124

15. Inverters..129
    15.1 Feeding inverters...130
    15.2 Isolated inverters...132
    15.3 Inverter technology...133
    15.4 Working principle of a pure sine inverter................................135
    15.5 Which inverter to buy...139

16. Overview: electric motors and generators...................142

16.1 Why is this a problem with photovoltaic installations?...........147
16.2 The working principle of heat pumps..................................148
16.3 Fridges and solar energy...149
16.4 Air conditioning and solar power......................................157

17. Circuit breakers and faulty currents..........................166

18. Lightning protection..176

19. Guerrilla photovoltaics...186

20. Suggestion for a small installation............................191

21. Forums and help..197

22. Orientation of the panels..200

1. Supplement: smart grid...206
The effects on the electricity suppliers:................................207
Implications to the customers..208

2. Supplement: night storage heaters..............................212

3. Supplement: measuring the consumption of appliances..........216

4. Supplement: comparing NASA, PVWatts and PVGIS.............218

5. Supplement: using a multimeter...................................220

6. Supplement: solar cooling...222
Fridge...222
Air conditioning..225
Cooling: can we do better?...227
Ice compartment and deep freezer...229
Modification of normal fridges and freezers........................230

7. Supplement: batteries 234
General information 234
How does a lead-acid battery actually work?......................236
General construction of a lead-acid-battery........................238
A few definitions..241

Crystals: how they grow and how they dissolve..................242
Sulfating..................244
Charging a battery and lead sulphate..................246
Acid stratification and the effects..................248
Corrosion in a battery..................253
Charging a battery..................254
Gassing..................257
Equalization..................258
Efficiency of batteries..................259
How to calculate the lifetime of batteries..................261
How much does a kWh from the battery cost?..................264
Rescue of batteries..................265
Battery interconnections..................267
Short summary of facts..................269

8. Supplement: battery management..................273

9. Supplement: electric fences..................276

10. Supplement: inverter for Minigrids..................280

Basics: the grid..................280
Devices in a Minigrid..................282
What exactly is a Minigrid..................282
Regulating a Minigrid..................284
The voltage in a Minigrid..................286
Switch box and switch-over box..................288
Requirements for inverters within a Minigrid..................288
Carrier Sense Multiple Access / Collision Detection..................290
Minigrids with three phases..................292

11. Supplement: some astronomy..................293

The orbit of the Earth around the Sun..................293
The time of day..................295
Some more astronomy..................295
The daily path of the Sun..................296
Global irradiation = direct irradiation + diffuse irradiation..................301
How to get an estimate of the diffuse irradiation..................302
Insufficient resolution of the irradiation data..................304
The course of the calculation..................304

Foreword

It has been quite a while since the last polymath lived. Since then it has become impossible for any one person to have profound knowledge in all areas, simply because there is far too much to know. It's not a stigma to know next to nothing in a lot of different fields of knowledge, in fact, it is perfectly normal for all people. You may have spent all your life in utter comfort without knowing very much at all about photovoltaics and there's nothing to stop you continuing your life without knowing more about that subject. However, something made you read this book; I hope it will help you to achieve whatever purpose you need it for. In this book we don't worry too much about details, we mostly deal with guesstimate figures, contexts and rules of thumb - in fact, everything necessary for a good orientation.

Zero calorie diets with full waitress service do not really make sense. The same situation applies to learning without effort. It is possible to reduce the complete knowledge on any subject to just what is essential to know on that subject, and then to transfer this essential knowledge with the help of examples which can be easily understood (you lose the exact detail, but then you can't have everything). That is the claim of this book about photovoltaics (photovoltaics is that part of solar energy which deals with the direct conversion of sunlight into electricity). On the other hand this book is ultimately a non-fiction technical book; don't expect that everything is cut in small pieces and prechewed. First a short explanation of why it might be interesting for you to read this book.

The first message of this book is:
to get a photovoltaic installation running is no big deal. The only thing you need to know is which way round to hold a screwdriver (well, you need a bit more, but not all that much).

The second message of this book is:
the necessary calculation needed to decide whether a solar installation pays for itself or not can be done (almost) on the back of an envelope.

The third message of this book is:
electric energy can be produced for **2.6 Euro cent** per kWh - but only when the sun is shining, and not everywhere in the world (but nearly everywhere).

Just by way of comparison: in Germany the typical householder pays 27 Euro cent per kWh, in Spain it is around 22 Euro cent, in France a bit less than 15 Euro cent. In Katar electricity is part of the basic services provided by the state and in wide parts of the US you pay a mere 7 to 10 Euro cent per kWh. Whether solar equipment pays for itself or not is not just a question of the local climate; there are other considerations as well.

That's why this book can't be more than a general guide. Whether some equipment might pay for itself or not is something you will have to calculate for yourself. If you want to avoid taking the risk that you make a wrong calculation, you can get in touch with companies that specialize in fitting solar installations, who will do all the necessary calculations and fit the equipment. Well, these companies do know the price of electricity charged by all the distributors and they know the price of all the necessary components. Most of these companies will calculate their price in such a way that fitting a photovoltaic system is only marginally profitable to you; the rest of the money is loot for them!

By chance I came across a web page with a calculator for solar installations, set up by a Dutch company. So I did the calculation using their calculator and then I did one - just for fun - using my own data. The price they asked assumed that the whole equipment would pay for itself within 15 years. From their estimate I subtracted the price of the materials, the amount that they would probably have to pay on wages and the taxes they probably would have to pay (I don't know the actual tax rates in Holland, so I just guessed). The result was that about 50% of the price was pure profit - after paying income tax! Without that profit the installation would pay for itself within 7 to 8 years in the not all that sunny country of Holland, and afterwards the owner could enjoy free energy for probably another 13 years! Wouldn't it be interesting to make that profit yourself?

You might now think that this profit (that the company makes) is really justified because they have had to learn over many years what to do and how to do it. At this point I have to admit that it really does take a long time to learn about photovoltaics and all the related topics. The reason why it takes so long is that everybody tells a different story, depending on his or her own interests. I have no particular interests (well, except that I want to sell as many copies of this book as possible), so I have the freedom to have a close look in any direction. Let's have a look who might be a target.

There are the electricity companies plus (at least in Germany) the owners of the high voltage distribution networks. Then we get the manufacturers of solar equipment, the commercial users of solar equipment, the politicians, and what are called in Germany Solateurs (they are the ones who mount the panels on to the roof), and a lot of other people. They all have their own interests in mind, they all tell a different story, but none of them tells me what is good for my purse (I suspect that they are all egoists!).

For about one year I dived into this topic and often it was quite useful that I'm a fully qualified electric/electronic engineer (7 years at university) - and I still do not know all the details! To read this book might take you a week. That means that I had to compress my actual knowledge by a factor of at least 50:1 and I had to give it a structure that all important areas are at least discussed in such a way that you can (at least roughly) understand what it is about. If you have a look at the sheer amount of Internet sites, books and brochures about photovoltaics, then it becomes clear that this book can't be more than just a short treatise, and that no matter how I'm going to do it, there will be people claiming that I'm wrong. Well, they are right, but this book here is still the one I would have loved to buy for myself a year ago.

Above all, I had to reduce the project. This book is just for private persons or the owners of small companies in all areas where photovoltaics already could pay for itself today (we will soon see which areas these are). In this book you will find no (further) information about subsidies and compensations. If you want to set up a commercial plant this book is not for you. If you bought a little chalet in the Andalusian mountains and you are not sure whether to pay for 500 meters of cable in order to get connected up to the grid or set up a solar installation, well, I think I have got here what you want to know!

No matter where you are on this world: if you think you get lots of sunshine and you cannot connect to the electricity grid, or they charge much too much money (that is, if you add up all the costs it comes to more than 15 Euro cent per kWh), you should take the trouble and read this book very carefully. The possible result might be that you save yourself a considerable amount of money or considerably enhance the quality of your life with little money.

What I can't do (and I don't want to do) is talk about particular devices or brands. What I can talk about are different types of devices. Let's just take batteries as an example. There are many types of batteries and they come in many different sizes. What I do is a short discourse about the most commonly encountered types and then I talk about the advantages and disadvantages of these types. I (hopefully) provide the necessary information so that you have the basic knowledge to make an informed selection when you later look through catalogues in order to find the parts you need for your installation. One purpose of this book is to increase the chances that this selection will be a good one.

NB All calculations were done on the bases of prices available on the market for normal consumers in 2013 - I must admit that sometimes I did just a little research. Since prices in the market change all the time, I left out all the peanuts in the calculations. When you pay back a loan, you normally pay interest only on the part not yet paid back. I used a short cut, and calculated only the interest rate on half the amount borrowed but over the full time. This is not absolutely correct, but is much easier (and faster) to calculate. I also used 20 years as time over which the loan is paid back and a 5% interest rate, just to make things easy and comparable.

Sometimes it is extremely difficult (up to impossible) to tell something technically correct and in a way that others can understand it. Please forgive me that I gave priority to readability. What is a technically absolutely correct book good for when nobody wants to read it?

In case you already know a bit about photovoltaics you will realize that at the beginning of the book quite a lot of facts are neglected which one should take into account. That was done on purpose so that readers with no previous knowledge don't drop the book in disgust after the first few pages.

1. Should this book be available in your own language?

So why not translate it yourself? The rules are very simple. On my site www.jdhenning.de there is a list of all the languages into which this book has been translated or is in the process of being translated. Let's assume you would like to translate it into Hindi and there is no entry in the list for this language. You then send me an e-mail with your qualifications and explain that you would like to translate the book into Hindi (NB there is no obligation for me to accept any particular translator). Then I send you a simple contract, in English, in order to regulate our collaboration.

The first thing this contract guarantees you is that you can take half a year to finish the translation and publish it as a book (that can be a print version or an e-book) with an ISBN-number. If you don't finish the job within half a year I have the right to terminate the contract between us and to accept a different translator for that language.

The book will be published in Europe and North America (in German and in English) and will be sold for 8.95 Euros or its equivalent as an e-book. For those who prefer to have a printed copy it is published as 'Print on Demand' for less than 20 Euros. You are free to set the price per copy to any price you like as long as it is not more than the equivalent of 8.95 / 20.00 Euros and you are free to set different prices in different countries if your language is spoken in more than one country. If you think that the people in a given country don't have enough money to be able to afford the book but need the knowledge urgently, you are even entitled to distribute free copies. I will retain 30% of the profit (income minus costs) and the other 70% you can keep for yourself. Making the translation is your contribution to the contract and does not form part of the costs. To me it does not matter whether you are a private person or a professional publisher. On the other hand, you must give me the right to check your book keeping and in case of big differences I retain the right to withdraw your right to publish this book.

What happens if there is a translation into your language but it is not possible to get a copy in your country - that could happen if a language is spoken in different countries but the book was published in only one or just a few countries? Why not become the publisher? I will arrange the contact to the translator and if he accepts you as a publisher, he will get 30% of your profit and I will get 30% of his profit. So the one who actually does most of the work gets most of the money.

If after a period of one year I neither receive any money nor any information why none is forthcoming, I retain the right to break the contract with the translator. Similarly, if a translator receives no money after a year from a publisher nor information why none has been received, he can break the contract with the publisher.

If I break the contract with a translator, the place of jurisdiction will be in Germany. If a translator wants to break the contract with a publisher, the place of jurisdiction will be in the country where the translator lives. I think these conditions are absolutely fair and simple.

2. Quick look at the German 'Energiewende'

We are in the year 2013. A few years ago the German nation (well, at least a big part of that nation) decided that it couldn't continue to live with the way energy was being produced, what with CO2 and grime, dust particles in our lungs and climate warming. It is simply not enough continuing to live and only just manage to survive, you need to live a lot longer and under acceptable conditions; if not, you are doing something really wrong.

Climate change does not allow you to survive the way the really crafty man survives in the following story:

> Two explorers are walking through the savanna. One of them asks: „What would you do if a lion were following us?" The other answers: „Well, I would put on my sports shoes." „But with sports shoes you will not be faster than the lion!" „That's not necessary. I just have to be faster than you!"

I will not personally experience the bad consequences of climate change (assuming it comes at all). Nor will all those other people who can already see their pension waiting for them at the end of the road. What bothers me is human stupidity! All the time the wrong questions are asked and then it is not all that surprising that the answers never help (if you pick your nose again and again you can't seriously expect that the result will be really different one fine day).

Roughly 40 years ago - after first detecting the ozone hole - it became common knowledge that the effect of human beings on the environment is not really very positive. Since that time a lot of very intelligent people brooded on the question of how the Earth could be preserved in a state which remains pleasant and makes survival worth the effort. To simplify: all the time it seems that the greedy win. Sales figures are much more important than the quality of life, especially if it is the quality of life of people living thousands of kilometers away.

The stupid thing is, that humans have always been like this since ancient times and it is not likely that we are going to change within the foreseeable future.

During the last few years, especially in Germany and other northern countries, more and more people at least tried to change their way of thinking. Something developed, which in German is called the 'Energiewende'. Very roughly the word means "a change of direction as far as energy is concerned" (did you think that German was always long-winded?). You might think about this as a fantastic or stupid idea; the important point is, that it got started - at least in Germany. The German government spent a lot of money (other people's money) and the project gathered speed. So much speed that it was threatening to get out of control.

Often you can read German newspaper articles explaining what has gone wrong with this project and the journalists (and not just German ones) complain without end. When I compare their findings with what went wrong at the new Berlin airport (to take just one actual example), I would say that the Energiewende project works nearly perfectly. Thinking of the number of well educated and capable would-be saboteurs, it is really almost a miracle that it works at all! But that is not the point now.

The point is that an artificial market was created. In the same way that wherever you find human beings you will find flies, demand for any commodity or service creates a market. That was perhaps a completely artificial market but it worked. Many smelled the possibility of making a quick buck and research projects arose like mushrooms - everybody wanted to build better and more efficient solar panels. Well, that part of the plan worked as intended.

Then came the big surprise. Everything developed much faster than expected; even in the wildest dreams of environmentalists the Energiewende was expected to take place some time in the distant future. Huge parts of the German population were amazed and started to set up their own alternative energy installations. They bought solar energy equipment and did not bother to calculate that it would never pay for itself. Competition on the market took off and prices dropped rapidly simply because solar panels started to be mass produced. But nobody really recognized this fact in time since German consumers simply bought up the whole world production of solar panels. They bought everything! At times there were more solar panels installed in Germany than in the rest of the world all together.

On the 21st of June 2013, at about noon, something really remarkable happened. More than 60% of the German electricity production was generated using alternative energy! Well, it was summer, but this result made it clear that full coverage with alternative energy is possible. In very simple terms the situation was that for full coverage in summer alternative energy supplies needed to be doubled, and for the winter multiplied by a factor of ten. Alternative energy was not a dream any more, to be realized in some distant future - it was at hand. No way could it be ignored any more. This situation had two effects:

1. The price for electricity at the exchange was often so low that generation using traditional power plants did not pay for itself any more (and that in Germany!).

2. The subsidies for alternative energy were cut drastically in order to give the owners of conventional power plants some measure of protection. The market for solar equipment collapsed and quite a lot of companies went bankrupt.

Now I slip into the role of a politician and ask myself some (previously agreed) questions.

Question: Could these developments have been foreseen?

Answer: Absolutely. All the relevant numbers had been published any number of times during the year.

Question: Isn't this to the disadvantage of the German economy?

Answer: If the management of a company is not willing or not capable of reading the writing on the wall, then the bankruptcy of these companies was only a question of time. The know-how is (not yet) lost! It was a necessary market adjustment! In addition, Germany's export surplus is much too high in any case.

Question: What, then, should be done differently?

Answer: The time for subsidizing the generation of electric energy is over (even if a lot of politicians and managers don't recognize that)! Either you are able to deliver electric energy or equipment to generate it for a market without subsidies (which is the free world market) or you see the red card and leave the sports ground.

Question: How is the situation going to develop?

Answer: Well, it was not just pure altruism that pushed the development of alternative energy in Germany. Part of the game was the hope that Germany would develop technologies that other countries don't have, in order to export them. If the German solar companies don't pull up their socks over the next three years then the German taxpayers will have financed the prevention of the world wide climate change without fair compensation.

Question: Isn't that a tremendous exaggeration?

Answer: No. Germany financed (almost) the complete technology needed for the Energiewende so that in other parts of the world electricity can be produced for just 2.6 Euro cent per kWh (that is, when the sun is shining). If this technology is used wherever there is sufficient sunshine, electricity can now be produced at an extremely low cost and without environmental side effects. The feared increase of pollution might simply not happen if solar equipment is used to generate electricity instead of just burning oil and gas. And it is self-financing!

Question: And why isn't all the world implementing this technology?

Answer: This is the One-Billion-Dollar-Question. I actually think it is caused through ignorance, laziness and avarice, or simply the interest-driven politics of those who will eventually lose. No, I have to correct myself: those who will - hopefully - eventually lose. For if they don't lose we will all lose.

Thanks for the conversation.

3. The costs of electricity

When talking about electricity prices it makes sense to first have a look at how this price is made up. I will give the figures for Germany, which will be probably quite different from those of other countries; for a start, German prices are higher than in any other country (well, except Hawaii and Denmark and few others). In Germany the domestic consumer has to pay 27 cents (all prices given in cents are Euro cents unless other wise stated) per kWh and they are made up as follows:

	Percentage	Cents per kWh
Production plus profit	34.0	9.18
Distribution and financial management	20.0	5.40
Renewable Energy Law	14.0	3.78
Electricity tax	8.0	2.16
Concession payments	7.0	1.89
VAT (Value Added /Sales Tax) 19%	16.0	4.32
Other fees	1.0	0.27

In Germany it costs 14.58 cents to generate electricity and distribute it to the consumer. In addition we pay 12.42 cents for taxes, concession payments and fees (in Germany the domestic consumer has to pay a fee so that big industry is freed from the payment of distribution costs).

If a person in Germany produces electricity and sells it then what he is doing is considered a commercial activity and he has to pay fees and taxes. That means that the price it costs him to produce one kWh must be lower than 9.18 cents. If his costs are higher he will not make any profit!

Now we compare prices in other countries with German ones. To get the numbers for countries in the European Union was fairly easy but for many other countries it is nearly impossible to find out exact numbers. The chart gives the electricity prices for households.

The next chart (source: Wikipedia; Railweh10) shows that even in the EU there are considerable differences in prices and the greediness of different governments is clearly visible as well.

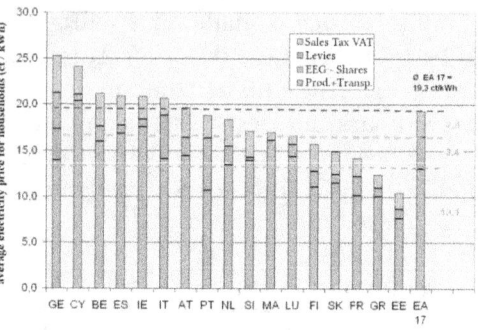

If you are living in a country that is not included in this chart (or the chart is really outdated because you are reading this book in the far future), then you will have to find your own figures. I have been told that there are some countries that make this just a little bit difficult. Often you have a base price, the cost of renting the electric meter or a fee depending on how many kW you might want to consume at least every so now and then. The best way to find out is by taking all the electricity bills you get over a complete year (if you had no connection to the grid, ask a friend) and add up each and any payment. This number you then divide by the number of consumed kWh and the result is the effective price per kWh (it varies from person to person). This value is important since it is the very base for knowing whether photovoltaics make sense for you or not.

3.1 The electricity price in Germany

In Germany there is the EEG law which regulates how alternative energy may be fed into the grid and how much the producer will receive per kWh. The basic idea of this law was (and is) to get away from the traditional ways of generating electric energy. The plan was based on a system of knock-on financing and small alternative power plants were to receive a higher price per kWh than bigger plants. Twenty years ago you could expect to get 40 cents per kWh for the next 20 years. In summer 2013 a small new solar power plant (just a few kWh) was receiving just a little over 14 cents per kWh.

When I read that for the first time I was a bit perplexed (I had not been interested in the financial aspect of PV before). If I have to pay 27 cents for every kWh of electricity I consume but only get 14 cents per kWh when selling it (and from that I even have to subtract fees and income tax) then I must be completely stupid if I don't design the installation in such a way that I myself will consume every erg of energy that I produce. If I can't produce as much as I need, I will have to buy electric energy. I would not want to feed anything into the grid (whether it makes economic sense to follow such a plan we will discuss a bit later).

It might in fact be a sensible idea to ask around and start a cooperative with the aim of putting together a slightly bigger installation. The aim of the cooperative is to deliver electric energy at as low a cost as possible to the members. Except for building up necessary reserve funds the cooperative does not need to make any profit, so there will be little or no tax to pay. Since the distribution must not be done over the public grid, there are no charges whatsoever. On the other hand you have to design and build your own distribution grid (I don't know about the legal situation in other countries, but in Germany such a cooperative is absolutely legal and there are even complete villages which did it this way).

Included in the 14 cents per kWh FIT (Feed In Tariff) was the profit. In Germany you will have to pay about 30% of the profit to the state as income tax. The idea behind the EEG law was not that the owners of alternative power plants could just cover their costs, the idea was to convince as many investors as possible to invest their money in solar panels and the only way to do that was to promise them a nice profit. For that to be possible the real price for generating electric energy with the help of solar panels and direct feed into the mains grid must be lower than 10 cents per kWh. - and that with the filthy weather here in Germany (at least during spring, autumn and winter)!

Now I was perplexed again. Was it really possible that I was the only one who looked at it in this way. That self consumption of alternative energy was the best legal investment one could do by saving 17 Cent with every single kWh produced! It took me some serious investigation to find out that I'm not alone. But I'm one of the very few who have no personal interest in NOT telling the whole story.

Lets have a look at who are the others and why they have no interest in letting people knowing this.

First there are the owners of the power plants. In Germany the situation is pretty simple: there are four big enterprises (with nuclear power plants and everything) and there are many municipal utilities with smaller power plants (often in the form of block heating power plants; in winter time people need more electricity and they need heat for their homes; the efficiency of block heating power plants is much higher than the efficiency of normal power plants since the surplus heat can be used).

None of the owners of the traditional power plants has the slightest interest in people starting to produce their own electricity. Additionally, it is a thorn in their flesh that they are forced to first buy the alternative energy before they are allowed to sell their own energy (the EEG-law forces them). That there have been big campaigns (such as "Energy supply not safe any more"; "Hundred thousand jobs in danger"; "Next winter there will be the super blackouts") is no surprise.

Then we have the owners of the super grids (very high voltage distribution networks). There are four of them as well (which is not just by chance). If electric energy is mostly consumed near to the place where it is generated, who needs them? Additionally there are the owners of the local grids and they have exactly the same motivation. And out of time long gone by (the proof would be difficult) they still have very good connections to each other and to the companies owning the big power plants.

Now we have a look at the state. Politicians want to be re-elected; if a big percentage of the population says that in the medium term future one should back out from the traditional way of producing electric energy (that includes topics like protection of the environment, global warming, and the typically German attitude towards nuclear power plants), then no politician can ignore this demand over a long time (at least he must not say so in public).

If something has to be changed, this will obviously cost money and it is no big problem to explain to the voter that the energy will cost more if the alternative energy has preference. This way round it works. But what would happen if loads of people started to build little power plants just in order to meet their own demand? Taxes worth billions of Euros would fade away (you remember that 12.42 cents in each kWh are for the state?). So, what to do? Right, you only subsidize those plants which feed into the grid. And since most people stop thinking when they can get something for nothing (or think they get something for nothing) this problem was solved to the entire satisfaction of the state.

This way of doing it has an additional and positive side effect as far as the state is concerned. Whoever wants to tap subsidies has to make a contract with the local supplier of electricity since they need to know from where and how much energy is going to come. Then the supplier can do all the rest of the administration as well and for the state there is no need to have additional personnel in the tax offices. And not a single cent less income for the state, because everybody is still using the same amount of electricity (this fact might explain why all the programs for reducing the energy consumption move along so slowly).

Then we have the owners of photovoltaic power plants. There are quite a few and they also have a powerful lobby. Those who owned a roof with a good enough orientation got the offer from the state: „You put solar panels on your roof and feed the electricity into the grid. The state guarantees you a profit of 15% for 20 years; an absolutely safe investment because the customer pays!" They did not need a lot of time to make their calculations. Some additional insurance costs (including damage and theft) lower the profit down to 10% but it is safe and guaranteed for 20 years. Just in order to make a little joke: this is called redistribution of income from bottom to top. And did they really want others to find out?

Last but not least there are the manufacturers. In order to understand their position one has to look at the fact that German customers bought up the complete world market for many years. Any year. The manufacturers of solar panels could not cope with the demand and the producer of the electronic components (inverters and so on) tried to improved the components for direct feed into the grid because this was what the market asked for. And additionally they developed the solar market for lifestyle products, interconnected high tech from the best (the only disadvantage was that the cost per kWh is nearer to 50 cents than to 27 cents). They are now crying for help because the subsidies were reduced drastically. Are they going to tell the world that they ran over the cliff because they only wanted to serve the high price segment? For sure they are not.

3.2 The situation today (end of 2013)

As already said, in June 2013 on midday 60% of the electric consumption in Germany was produced by alternative methods. The Energiewende did not just make a start, it won all along the line, worldwide. You are not convinced? We have the proof that all components for solar energy production can be build as mass products! Electric energy can already be produced today for 2.6 Euro cents per kWh. We only have to drop the idea that solar plants have to support the grids instead of working locally. However, there will be some consequences (first for Germany and later for others as well).

One consequence will be that the dirt belchers of the nation will have seen their last days. Brown coal power plants will not be needed any more; they can run as long as they need no maintenance or new investment but then they will stop for good. The same applies to the black coal and nuclear power plants. It took about 20 years to climb the steep and stony part of the Energiewende path but it will take less time to finish it (that is why all the belchers were running like mad in 2013; the investors wanted their money back and compensation for profit not gained in the future is in their eyes definitely adequate).

In the very near future we will all know whether this adventure in environmental politics pays for itself. Whether it is really possible to do something good for the environment and in addition save money. Whether alternative energy is really apt to have strong positive effects on the economy of those countries which start an Energiewende themselves (or at least not hinder this development) will be seen at the latest in 10 to 20 years.

Or in order to make it short: if I had not come to the conclusion that the worldwide energy revolution makes sense, can be done and will pay for itself, I would not have written this book. As a German engineer I learned something really well: to calculate! And your job is to check what I tell, because only if you checked it yourself can you really be convinced. And that is the first and essential step for engaging first gear.

So, let's get moving!

4. A minimum of physics

I think we have to tackle this topic first, after which you are finished with it. The rest of the book will not be harder than these next three pages. After climbing over this (psychological) hill the complexity factor only goes downhill, with here and there perhaps a few small hillocks.

Imagine you have a bucket with a hose connected near to the bottom; at the other end of the hose is a tap. Now we fill the bucket with water and put it on the table; the tap rests on the ground. If we open the tap a very small amount of water will flow out with only a little force.

Now we lift the bucket, using a rope, to the first floor and put it on a window sill. If we now open the tap (still resting on the ground) the water comes out with much more force. We take a measuring jug and find out that after 10 seconds (for example) half a liter has come out.

Now we lift the bucket to the second floor and put it on another window sill and we repeat this little experiment. The pressure is much higher (to be precise, it will be twice as high) and after 10 seconds we have a complete liter in our jug.

It was exactly this experiment that Herr Ohm carried out about two hundred years ago, except that he was not using water but electricity. The height above ground at which we placed the bucket is now represented by the electric tension; this is the force with which the electrons are moved. The electric tension is given in Volts (the formula symbol in English speaking countries is a capital letter 'V'). The amount of water per second equals the number of electrons per second moving through an electric conductor. The number of electrons per unit time that move along the conductor is given in Ampere (whose formula symbol is the capital letter 'I').

Now just one thing is missing, namely the electric resistance. This we could imagine as the narrowness of the hose or as the resistance to the flow (and the formula symbol is a capital 'R'). If we halve the cross-sectional area of the hose then only half the amount of water can flow through the hose, and the resistance is doubled.

What Herr Ohm found out is simply this: if, with a constant diameter of hose, the drop height is doubled, then the amount of water flowing through the hose doubles as well. When the drop height is tripled the amount of water running per second through the hose triples as well. If the cross section area is halved then the amount of water coming out is halved as well. And this is how he formulated what later came to be known as Ohm's Law (mind you, at that time he first made that law it did not have that name yet; the * symbol means "times"):

    Drop height = flow resistance * amount of water per unit time

or

    Tension = electrical Resistance * Current

or using the customary formula symbols (V = Tension in Volt, R = Resistance in Ohm, I = Current in Ampere)

    $V = R * I$ (or V equals R times I)

If the size of two components of the formula are known, the third one can be calculated (always!).

Now we have to add something and that is power. Power or effort as we usually call it; I think that it really took me some effort to pull the bucket twenty times up to the second floor.

It is clear that it takes twice as much power to lift the bucket up to the second floor instead of the first floor in a given time. And if there is more water in the bucket it takes more power as well. Power is (not only with electric current) measured in Watts. Since the necessary power increases linearly with the drop-height and with the amount of water, we get the following formula:

    Power = Drop-height * Amount of water per unit time

or

    $P = V * I$ (or P equals V times I)

The two equations are related because current (I) is defined as number of electrons per second, so the time unit is already included. There are two further variations of the second formula. If we take $V = R * I$ (the first formula) and solve it for I or V and insert that equation into $P = V * I$ (the second formula) then we get additionally:

and
$$P = I^2 * R \qquad \text{(or P equals I times I times R)}$$
$$P = V^2 / R \qquad \text{(or P equals V times V divided by R)}$$

Well, we are over the top of the hill. What we still have to do is to develop a feeling for what we have just learned, what it means in your daily life (don't tell me that these formula never showed up in your daily life; they did but you simply did not recognize them). And we have to deal shortly with the two sorts of electric current.

Lets take a desk lamp with a 75 Watt tungsten light bulb. From the mains comes electricity (at least in most parts of Europe) at 230 Volts. We know the power and the tension, so we can calculate the current which must be flowing (two factors determine the third one). We divide 75 Watts by 230 Volts to get 0.326 Ampere (75 W / 230 V = 0.326 A). Such a calculation we might need to make in order to find out what size fuse we might need.

Or we have a car with 75 kW (kW means kilowatt where kilo means a thousand; the car has the power of 75,000 Watts which, by the way, equals 100 Horse Power or HP). Power is work done per unit time. If I want to know how much work the car delivered then I have to multiply the power by the time (delivering 75 kW **power** during one hour gives us **work** of 75 kWh; kWh is a kilo Watt hour). Whether you consume 75 kWh for one hour or 37.5 kWh for two hours makes no difference; the total amount of work done comes to 75 kWh.

Or it is winter time and we put on a little electric heater in our living room because it's so cold in there. We have a look on the label of the heater and see that it is rated at 2 kW. A horse can deliver a steady 750 Watts (it can deliver more, but only for a short time). If electric energy were produced using horses, 2.66 horses would have to toil so that we get warm feet.

I think that all was not too complicated; we need these formula later in order to calculate the size of our solar installation.

Now we come to the last part of physics for photovoltaic installations: different kinds of currents. If you read this chapter carefully you might ask right now how that could be possible if the electric current is simply defined as the number of electrons per second moving along a wire. Right, it depends on the way they move.

In our example with the buckets, the water flows downhill all the time. With electric current this is called DC as in Direct Current. DC comes, for example, from batteries.

DC used locally is very practical but causes problems when it has to be transported over longer distances. Any cable has some resistance and this resistance produces transmission losses (i.e. the cable gets warm); these losses can be calculated using the formula $P = I^2 * R$. The losses increase with the square of the current. The formula $P = V * I$ tells us that there is only one way to keep the losses low, assuming we use the same size cable. If the current must be made smaller then we will have to increase the voltage. This is now the problem with DC - there is no cheap way of changing the voltage up or down at a transit station. Furthermore, using high voltage DC in the house is not a very good solution either because you would have to keep half a meter (or more) away from all sockets, for safety reasons.

The very first (local) grids were laid using DC. Then came a challenger called AC (or Alternating Current). Current is the flow of electrons; in a DC circuit the electrons run all the time in the same direction, whereas in an AC circuit they are moving back and forth. In Europe the electrons reverse their direction 100 times per second, that is, the frequency is 50 Hz. Many electric devices do not care at all whether they are connected to DC or AC as long as the voltage is suitable.

However, there is a funny effect with AC since the electrons have to stop for a very short moment when they change direction. A normal tungsten light bulb shines because the electrons collide with the atoms of the filament, which heats up the metal until it is glowing white. When the electrons change direction this heat production stops and for a very short period there is no light. Humans don't notice this effect because their eyes react too slowly, but think of some ETs with different eyes coming to visit Europe at night. The whole continent gets illuminated 100 times per second. These Earthlings must be crazy!

The big advantage of AC is that the voltage can easily be changed at a transit station. The only thing one needs is a transformer (ultimately a transformer consists of two copper wires which are wound around an iron core; that's all and it is relatively cheap; big ones have efficiencies of 99% and more). That makes it possible to have a very high tension distribution grid for transporting electrical energy over hundreds of kilometers, a middle tension grid for delivery in a smaller region and low tension mains which delivers the electric energy into the house (in most of Europe with 230 Volts).

The disadvantage of AC is that it is not possible to store the energy directly. It first has to be changed into DC in order to charge batteries or it has to be converted for some other kind of energy storage. We will have a closer look at this a bit later.

Well, the three pages with physics are finished and I think it was not really difficult to follow my explanation. If you did not understand everything that's no big problem for the moment; just read these pages every now and again and understanding will come. If you really did not understand anything, look for somebody who is better in explaining it all. Read the rest of this book if you like, but don't try to set up photovoltaic equipment without the help of somebody who really has competence concerning electricity.

5. Costs running a generator

In northern Europe an average household of two adults and two children will use about 10 kWh per day. Lets assume there is such a household which has to or wants to live without a connection to the mains grid. When you are living in an energy-aware style, your energy consumption can be reduced to about 5 kWh per 24 hours without any harsh cutbacks in your life style. This average household would be living in a properly insulated house, with a gas-fired central heating system (so we are actually not talking about the tropics) that also provides warm water, a washing machine with cold and hot water feeds, the washing dried in the garden or the loft, dish washing is done at the sink after the meal, anything is switched off if not actually being used. Generally they could live a perfectly normal life without having to watch every erg.

But where can they get the 5 kWh per day? Well, we could buy a super duper generator, one that is large enough to power the whole house when run at its normal level. Depending on which country we live in, we can either buy a generator that runs on petrol and convert it to run on LPG or some other type of gas, or we can use a diesel generator. Different countries have different fuel tax rates, and what might be cheaper in one country might be more expensive in another. I will be using numbers that are typical for the EU but be prepared to plug in your own numbers to see what it would cost you. So we will calculate the liter of petrol to cost €1.50 and its equivalent amount of butane gas would cost €0.80.

So we invest the odd €500 in a 2.2 kW generator - you can get them for as little as €200, but this more expensive machine is supposed to be a really quiet one and it should last much longer. A quiet generator, with a maximum noise level of 65dB, should not make much more noise than a washing machine when spinning (the neighbors will not jump out of their bed because you started to produce electricity). The generator will easily run a modern "slow-start" washing machine, although perhaps you had better not switch on your 135 cm plasma TV or microwave till the spin drier has stopped (that are the little limitations you would have to live with).

You switch on the generator first thing in the morning, and switch it off late at night before you go to bed - that's 16 hours a day (during the night you had better use a lighting system that runs on batteries and you need a similar solution for the fridge and the freezer). In the course of a year that little generator runs 5,840 hours. And after only two years the generator starts acting up - small generators have a bad name in general.

The reason for this is almost always "lack of maintenance". Look, many of us have a car these days; on average we drive (in many parts of Europe) about 400 km per week. Even out in the country you will be lucky to average much more than 70 km per hour, which means that you drive about 6 hours a week. Per year you spend 300 hours in the car and drive 20,000 km; after such a period of time the car needs a service and most cars these days need new filters, an oil change and perhaps a little adjustment of tappets or valves. If the generator is used for 16 hours a day, it was running these 300 hours within 20 days! Oh, now do you see?

Using a generator means that it should be serviced about every 1,000 hours (i.e. at intervals that are three times as long as those for a car), which in practice means about every 2 months. A service is no big deal; something any knowledgeable handyman can do in less than half an hour. If you cannot (or don't want to) do this yourself, you should set aside a certain amount of cash to get this service carried out 6 times a year. Only a few years ago it was generally held that the motor of a simple car was good for about 100,000 km; so don't expect the generator to last much beyond five years.

The last thing we have to find out is how much fuel such a generator uses to produce 5 kWh per day, with now and then a full load, and a lot of idling in between. A 2 kW generator typically uses 1.5 liter per hour under a full load; when powering just a dozen low-energy bulbs and a small music system it will need 0.5 liters. Without any rigorous proof we can guesstimate that this comes to around 13 liters of unleaded petrol per day, which comes to about €20. That works out at about **€4.00 per kWh** instead of the €0.27 per kWh that the (German) electricity company might charge. Not a very good deal.

Now most of that cost is in fact the cost of keeping the generator running for 16 hours a day though it is producing a mere 100 watt or less most of the time. Suppose we use the generator just a few hours, simply to charge a few large batteries and possibly run the washing machine, and for the rest of the time run our lights and TV from batteries.

In a 100 Ah 12 Volt battery we can store 1.2 kWh. We actually need 5 kWh in the course of the day; if we include a small safety margin we will need 6 batteries with a total capacity of 7.2 kWh. We also need a good battery charger which will cost about €350. This will enable us to charge the batteries within 3 hours (we would be torturing the battery but I neglect that for the moment) and that is now the time the generator needs to run every day. We are generous and have it serviced twice a year. Without idling, our fuel consumption will be much lower (3 hours a day * 365 days * 1.5 liters * €1.50 = €2,460). We assume, that the components last 10 years.

Generator	€500 / 10	50.00 €
Batteries	€600 / 10	60.00 €
Battery charger	€350 / 10	35.00 €
Inverter	€250 / 10	25.00 €
Service	€20 * 2	40.00 €
Fuel		2,460.00 €
Total per year		2,670.00 €

Now we come to a cost of **€1.46 per kWh**. This is a lot better than the €4.10 in our last attempt but we are still far above the €0.27 that the electricity company would charge. And it is obvious that it is the fuel costs which are the biggest component.

We try to use cheaper fuel. In many countries you have fuel for cars and lorries with a relatively high tax on the fuel and fuel for heating the house or powering stationary motors with less tax on it (this could be diesel or butane gas etc.). The normal petrol costs €1.50 per liter; the energy equivalent in butane costs just about 80 cents, but we need to use a conversion kit which will cost us €200, although we only need to buy that once.

Generator + kit	€700 / 10	70.00 €
Batteries	€600 / 10	60.00 €
Battery charger	€350 / 10	35.00 €
Inverter	€250 / 10	25.00 €
Service	€20 * 2	40.00 €
Fuel		1,312.00 €
Total per year		1,542.00 €

We now spend a mere €1,542 per year but this still comes to **€0.85 per kWh**. If you don't have a source of much cheaper fuel this is as far as we can go, because in the calculation I have even left out the cost of financing, not to speak of a shed in which to put the generator and the weekly drive to the gas station and all the rest. Absolutely no chance of getting it cheaper within the EU!

As you see, running a generator is an expensive hobby. However, by and large, running your own generator is not going to be cheaper than buying electricity from the electricity company.

Where to find cheaper fuel? Unless you have a brook running through your garden, hydro-electric power is not an option. Nuclear power will be far out of reach, if desired at all. Coal, black or brown, is no option either. Before you get it delivered to your doorstep it is already that expensive (transport, taxes, intermediate dealers etc.) that you can't compete with the electricity companies. No way, no chance, not even remotely!

If, after some months of investigation, I'm not completely mistaken, there is only one option left to generate (hopefully cheap) electricity for your own needs: solar power! Either in the form of wind turbines or by photovoltaics.

As the title of the book indicates, we are not going to talk about wind turbines. We are just going to check how cheap electricity produced by using photovoltaic panels could be. As always in this book, it is an guesstimate. I only want to get the correct range for the price.

Remark: the situation in industry is a bit different. Some decades ago it was perfectly normal for a factory to have its own power station. They could buy coal, gas or other fuel for cheap prices since they bought bigger quantities. And their purchase was (more or less) tax-free (the money they paid for the fuel was subtracted from their receipts before these were taxed). In the course of the popular 'outsourcing' movement these power plants disappeared. During the last few years they reappeared in the form of wind turbines and photovoltaic installations; it looks as if industry already knows where to save money (beginning of 2014 the German Government already put these power plants on their target list as tax sources and industry is talking about going to the institutional court; that will become very interesting).

6. Its possible: one kWh for 2.6 Euro cents!

In the Internet I found an interesting article[1] published in October 2012. It states: „For example, central inverters are available in India today at about $0.12 - $0.14/Wp, string inverters cost about $0.18/Wp."

1) http://www.re-solve.in/perspectives-and-insights/indian-solar-industrys-love-affair-with-central-inverters/

Now first you have to know what 'Wp' means. Wp stands for Watt peak. An inverter is a device for transforming DC into AC, for example from the 12 Volt DC produced by the car battery to 230 Volt AC for running a standard mains fridge. A central inverter is a pretty big device and is good for a few hundred kW up to the Megawatt level. String inverters are relatively small devices for a few kW each but it is very easy to connect hundreds of them in parallel in order to reach higher power levels. If we convert from Dollars to Euros the price for buying a string inverter (and they are the only ones we are interested in) will be roughly 150 Euros per kW inverter power (sine wave).

On a German website[2] solar panels were offered on the 11th of June 2013 for 0.61 Euros per Wp (sales and import tax already included). Efficiency of the module was given as 15%; the price per square meter worked out at 87.84 €.

2) http://pv-vertrieb.biz/module.php

Now I have to deal with a little problem concerning the calculation of photovoltaic installations: there are almost no worldwide irradiation data available free of charge - everybody who knows something seems to want to get paid for that information. I found just one free source for irradiation data (and even after quite a lot of investigation not a single other one) and that is a database set up by NASA. The problem is that these data were not measured directly so they can't be considered really trustworthy. I cross-checked the data given in the NASA data base with probably more reliably measured data published elsewhere (see supplement No 4). The average over the year coincided well but I saw differences of more than 20%, especially during the winter period.

Then there are some data from PVWatts, Version 1. This is also an American data base; so there are a lot of data sets for the US and only a very limited number for places outside the US. For all other data you have to pay. Originally I wanted to select for my example of how to calculate a price (we'll come to that in a moment) a place in Mexico but PVWatts has no free data for Mexico. So I took the data from Amarillo (Texas) but used the labor costs prevalent in Mexico (I don't want to imply that there might be illegal Mexican workers in Texas; really not).

We take the irradiation data of Amarillo (Texas) (35° North; PVWatts, Version 1, standard settings) at 5.86 kWh per day per square meter. With an assumed efficiency of 15% we are able to produce 0.879 kWh per day from a square meter of solar panel. In order to generate one kWh we need 1.14 square meters of panel which comes to about 100 Euros. If we want to deliver 1 kW every hour for ten hours every day, we need ten times as many panels.

That gives us the following estimate:

Inverter	150 €
Solar panels	1,000 €
Installation (labor)	100 €
Total	1,250 €

We finance the equipment with a loan over 20 years and at a 5% interest rate.

Back payment per day	17.12 cents
Interest rate per day	8.56 cents
Daily payment	25.68 cents

Since we produce 10 kWh per day we get a price of 2.57 cents per kWh electric energy. The real costs might be a bit higher but my point is that this is a reasonably realistic estimate. We can make further calculations using a figure of **2.6 Cent per kWh** in solar electricity production.

Here we have proof that electric energy produced using solar panels is cheaper than electricity from coal power plants or even nuclear power plants. There are no distribution costs and there are no taxes or fees to be paid - I expect that in the future quite a few governments will become very creative in how to tax solar panels (stop the press, I have new information: in the summer of 2013 already one state started to be inventive; see the next chapter).

But I have to admit two things:

- This way of generating very cheap electricity works only in the zone between 37° North and 37° South. Just to name some cities: on 37° North you find San Francisco or Málaga, on 37° South Buenos Aires or Melbourne.

- Cheap electricity is only available when the sun is shining. For spoilt people from northern Europe or from the US this might be a shock (no light in the night) but for somebody who has had to live without any electricity at all this offer is more than half of what he needs. Also, there is the possibility of using batteries but then the electricity will not be so cheap any more (details follow later).

We have a closer look at the question for whom photovoltaics of this type might be interesting and we will see that the 37°N to 37°S limit was set (a bit) arbitrarily but not completely without reason. And we will see that photovoltaics can be sensible outside this belt as well.

But first we have a look at how the big energy companies start to fight back.

7. Drunk, stupid - or simply criminal?

Did you ever receive a bill from a restaurant for not eating there, though you could have done? Probably not. Last night, I heated up a tin with soup, absolutely legal as I thought, and the next day I received invoices from all the restaurants near by claiming amounts between 4 and 50 Euros. I would not even consider paying a single cent to any of them. But the Spanish Government is actually passing a law (end of 2013) for exactly that. You don't believe me? Well, in that case, read on!

Spain, together with Portugal, Italy and Greece, is one of those countries in western Europe where photovoltaic systems will make most sense. Actually, when taking a good look, it is hard to understand why they are not used more intensely. Spain has to import 80% of it's energy needs and has to pay, year in, year out, 40 billion Euros for these imports. Under these circumstances one could suppose that the government has a big interest in changing this situation because any kWh from photovoltaics lowers the export deficit.

But on the contrary; those who set up a solar installation, at their own risk and costs, get called freeloaders. That is why I selected such a harsh title for this chapter. It is actually near to unbelievable what is happening in Spain.

7.1 First of all: what's going on?

In 2011 a law was passed which legalized the installation of solar equipment for personal use (autoconsumo). There were only few people who decided to get their own solar power plant; they ordered solar panels and covered their roof. Then, at the beginning of 2013, the government stopped paying the Feed In Tariff (if you suspect that this was some kind of breach of contract, you are probably right).

In July 2013 the government decided to make a law which introduced a so called 'fall back levy'. This law will most probably be passed by the Spanish parliament in spring 2014. If somebody owns a photovoltaic installation and also is connected to the grid, then he will switch over to the mains grid (for which he is paying as well) when his solar equipment slows down for the night (or bad weather). For permission to use this kind of additional power supply, the owner has to pay a fee of 6.2 Cents for any kWh generated by his solar equipment. The time to amortize the equipment would go up from between 12 and 15 years to far more than 20 years. With such prospects all investors will turn away in disgust.

Somehow I must have misunderstood something completely about free markets and capitalism. If my memory serves me correctly, the word competition was involved: whoever makes the better offer will get the buck!

Lets say, I walk through a Sky supermarket and a bit later I visit an Aldi market (two big chains in Germany). Then I have a look into my purse, after which I decide that I like the prices and the quality in the Aldi supermarket much better; so I do my shopping there. Just as I arrive at my home with my plastic bags, somebody from the Chamber of Commerce rings at the door and claims €5.50 from me. The Sky market is forced to keep its infrastructure going so I must pay them compensation. Or bicycle riders have to pay an extra fee for the preservation of petrol stations. Or all adults have to go once per year to a medical examination of their lungs subject to extra costs; if the inside of the lung is not black the citizen will face a trial because of tobacco tax evasion.

I hope you get the idea; I tried to turn the story into an absurdity but that is hard work. A friend of mine told me years ago: „There are three enemies that you should never try to fight: secret services, Mafia and the tax office!" I would really be surprised if any secret service were involved in Spanish energy prices (at least directly). Two possible culprits are left. That starts to be interesting!

Now you might object that Spain is a sovereign state and that they can make up new laws as their parliament (or whoever) decides that are needed. I fully accept that but on the other hand, Spain does have a constitution and it is bound (at least a little bit) by international treaties (and if Spain does not fulfill them, then the other countries might take legal action).

In all constitutions that I know of, property is guaranteed. Part of that is to ensure the preservation of the status quo; if I legally own something (and that could also be an entitlement to certain rights) then this may not be changed by a subsequent law. There was a time when it was absolutely legal to buy solar panels and to sign a contract which guaranteed a price for feeding into the grid. Now the Spanish government says that they have decided unilaterally to cancel all these contracts and that from now on, contrary to how these contracts were originally drawn up, the owners of photovoltaic installations will have to pay a fee of 6.2 cents per kWh for feeding into the grid. That is a breach of the constitution, even in Spain. No possible doubt remains! I read in an article that even Spanish lawyers were taking their solar installations down, which they themselves must have felt was the only sensible decision.

The official reason for all this is that because of the Feed In Tariff the state accumulated a deficit of 24 billion Euros and this amount increased by 4 billion every year. The people who caused this debt, these parasites of the electrical system, must repay the debt. Knowing that photovoltaic systems only deliver 3% of the total electric energy in Spain, this argument does not look all that convincing. So I dug deeper and found a very interesting statistic[1]. In Spain, in 1999, generating electricity cost roughly 4 cents per kWh and its distribution was costing 3 cents. In 2009 generating electricity cost roughly 4 cents per kWh but its distribution was costing 6 cents.

1) http://www.e3analytics.ca/wp-content/uploads/2012/05/Analytical_Brief_Vol4_Issue1.pdf

Spain consumes per year 267 Terawatt electricity, tor, written in full, 267.000.000.000 kWh. Somebody organized matters in such a fashion that distribution was costing 3 cents more per kWh in 2009 than in 1999. That amounts to some 8 billion Euros per year; that is not tax paid to the state but money the state pays to somebody! Then we take the additional 4 billion Euros on debts the state generated for distribution, and we get nice little 12 billion Euros that somebody put into his pockets. And please note, that amount is pure profit, since the Spanish grid did not improve all that much during this last decade.

Oh, before I forget to mention it: in Spain the government decides about changes in the grids and the prices to be paid. Now we can be pretty sure that the tax office is not the wrongdoer! Who is left?

7.2 My home is my castle!

Privacy is in many countries held in high esteem (if we leave out the British and American governments; to them the privacy of others seems worth nothing). In any civilized country even the police has no right to enter my home and there are only very limited exceptions to that. The first exception is an empowerment given by a judge (normally called a search warrant). The second exception is given when there is actually a threat to life or the possibility of physical harm. The third exception is given when goods of high commercial value are in danger and the breach of privacy is absolutely minimal (so firefighters may cross your lawn when the house of the neighbor is in flames). And the last exception is when the safety of the state is in danger. Fine!

What does the new Spanish law stipulate? Firstly that you have to connect your photovoltaic system directly to the grid (if available) so that the number of kWh you generate can be metered. The actual interpretation of the law to come is that you have to feed in what you are not using up yourself. Not only do you not get any compensation for that but you have to pay the 'fall back fee' of **6.2 Cents for any and all kWh you generate.**

What happens if I secretly generate my solar energy? Most probable that will be a very risky undertaking since the new law stipulates a fine of between 2 million and 30 million Euros (yes, you were reading that correctly; the minimum fine will be 2 million Euros!). So how will they find out? Very simply; the inspectors of the Ministry of Economy, Tourism and Energy are given the right to inspect any plot, house or flat if they have any suspicion that somebody is generating his own electric energy, without even showing that they have grounds for that suspicion. Very well, I simply don't let them in! Good idea but it will not work out, because the inspector simply drives to the next district court and gets a search warrant from the judge straight away. Then he comes back together with the state police. If you don't let them in, they can, and will, break down your entrance door.

So what happens after this? I think you will go straight to jail. If two million Euros is the minimal fine, the chances are astronomically high that you will try to run away. Not just foreigners, but everybody. The use of unapproved solar installations will be treated as a major offense! In the end it will have proved to be useful to the Spanish State that they did not forget how the Spanish Inquisition was organized and how well it worked.

You don't wish to believe all this? Well, that's your right. In Germany we have freedom of religion and you are free to believe whatever you like; even if you believe in the big Spaghetti Monster who created the world. That is why I prefer to stick to the facts[1] (sorry, but it's in German)!

1) http://www.heise.de/tp/blogs/2/154673

So here we have the first open attack on solar panels in private ownership and you really don't need a lot of fantasy in order to locate the culprits. However, privately owned solar panels are not their only victim, since in fact they even (almost) killed a very big project.

7.3 Desertec

The very basis of this idea came from the so-called Club of Rome, who felt that it should be possible to use the power of the sun in the Sahara Desert and transmit the electricity using high-voltage DC links to Europe. Many people smiled at such a project because investments of hundreds of billions would be necessary. But technically it could be done!

They planned to use thermal solar power stations. Very long curved mirrors would be used to concentrate the rays of the sun on to a tube (the mirrors can be moved so that the rays of the sun focus on the tube all day long) through which a special oil would circulate. Part of this oil would go through a phase change storage unit and the other part would vaporize water; the vapor would drive a turbine and the turbine would turn an electric generator, thus producing electricity. In the phase change storage unit a special kind of salt is stored, the temperature of which is near to its melting point. A gigantic amount of heat can be stored by melting the salt and this energy can be recuperated when the salt becomes solid again (the nice thing with these storages is, that the temperature does not change as long as there is liquid and solid salt). In this way the production of electric energy could continue day and night. Such storage would even allow electricity to continue to be generated even if there were a few days without a lot of sunshine.

All the necessary components of such power plants have been known for decades and have been tested extensively. Such electricity would cost 6.5 cents in the desert. If high-voltage DC links are used, their efficiency can be so high that only a loss of 3% per thousand kilometers will occur;

that means a loss of 10% on transmission from the Sahara Desert to the north of Germany. That is easy to live with. Take it from me (or go back to chapter 3 if you want to check) that the price for conventionally generated electricity is 9.18 cent per kWh (including profit). So desert energy is competitive!

A nuclear power plant delivers that same kWh for 3 to 4 cents, which is a lot cheaper. On the other hand, energy prices go up every year by something like 3%. It would not have taken an unacceptable period of time until the price of desert electricity would really be interesting. It looked as if the project could start up when some big German enterprises (insurance companies, banks etc.) smiled at the idea and started to engage with a show of enthusiasm. These companies founded the Desertec Industrial Initiative (Dii); one member of the Dii was that very same Desertec Foundation who launched the original idea.

On the 27th of June 2013 the Süddeutsche Zeitung (a relatively big German newspaper) published a report that the Desertec Foundation had stopped their cooperation with Dii. The official reason given was that two members of the board fell out with each other and that there were some difficulties with Spain. The first reason is not very creditable, since usually investors become pretty unscrupulous whenever their money (or a profit in the near future) is at risk. In less than 24 hours two other managers would have taken over. Full stop.

That there had been problems with Spain sounds much more probable and was confirmed within the next few weeks. The Spanish Government refused to allow the high voltage DC links to cross Spanish territory. Normally, being a German Energiewende fanatic, I would feel the need to complain about that, but when I did the research for this book I unearthed the fact that a zombie had been smitten. This would also explain the troubles in the board of Desertec and why the managers were not replaced. There simply was no reason to replace them.

If you have a look at the climate data of the Sahara and compare them with the data from Spain, Italy or Greece, then you can calculate how many kWh you could generate there and what they would cost you. The result - surprisingly - is that it is cheaper to produce electricity in southern Italy than to produce it in Africa and then transport the energy to Italy (slightly, but the effect is there). What is more, the political risks are very much smaller.

Since no German enterprise is going to pay for the well being of the people in Morocco or Libya there was simply no base for starting that industry there any more. So they were not even angry with Spain since they had somebody else to blame. In which case, why not run the dirt belchers a little longer and let them earn some more money.

8. Photovoltaics and the seasons

There are areas in the world where people would be very much happier if they had a little less sunshine. On the other hand there are also areas where you don't see the sun at all over long periods and it is bitterly cold (for example north of the northern polar circle). In this area you have at least one day during which you don't see the sun at all (or not completely) and for compensation there is also a day on which you see the sun all day long (or more than half of it).

In order to supply somebody living within the arctic circle with solar electricity, it would be necessary to collect a huge amount of energy during the short summer and store it for the long winter time. To supply a polar station with electricity from solar collectors would, in theory, be possible but there are many other ways to supply the station with much cheaper energy.

The other extreme to the polar regions are the tropics. Between the northern and the southern tropic (23.5° North and 23.5° South) things are completely different. Every day is 12 hours long (well, not exactly, it does vary slightly). At six o'clock in the morning the sun rises and it sets at six in the afternoon. For all practical purposes, one can neglect the difference in length of solar days.

If you live in this region, and you want to store energy for the night, then the night is 12 hours long and to store energy for such a period is relatively simple (we'll look a bit later at the actual cost of storing energy). Just in order to give you the motivation to read on: to store one kWh in a simple battery costs 10 Cents; this is not really cheap but often cheap enough to be competitive!

Depending on the latitude there are noticeable differences as far as photovoltaic installations are concerned. Outside the polar circles photovoltaics don't make sense at all and in the tropics photovoltaics make a lot of sense. So we can be sure that there is a big gray area between the tropics and the polar circles of varying suitability for using solar energy.

We already saw that Germany took over some kind of pioneering task in photovoltaics, and, based on it's economic power, it had the resources to do so. But was it a reasonable decision? As a technician I would say it was downright stupid! On this planet there are only a few habitable areas where photovoltaics make less sense than in Germany.

This might sound a bit crazy looking at the results but you can convince yourself easily. Just take a globe (the sort with a map of the world on it) or a normal map giving latitudes. Now you take a look at Munich (a big city in the south of Germany; known worldwide for its big beer party in October) and hold you finger just over it without actually touching the globe. Now keep your finger in that position and turn the globe until the American continent is under that finger. Where does it point to? Yes, that's right, the middle of Canada! Now we go back to Germany and you hold your finger over Hamburg (in the north of Germany); now turn the globe once more until your finger hits the American west coast. Where does your finger point to? Right, the south of Alaska!

Nobody with at least a small amount of common sense would want to go to Alaska in order to set up solar equipment (a former German politician once promised to become a pineapple farmer there but he did not keep his promise; pity). However, the Germans did - no, not pineapples in Alaska but photovoltaics in Hamburg or even north of it. Are all Germans completely crack-brained?

Well, the professional world is still discussing this topic! My personal opinion is that, after having taken a good look at all these big and influential groups who don't like the idea of alternative energy at all, we have in fact succeeded pretty well. On the other hand, if we don't want to leave our offspring (personally speaking, I don't have any so I'm not really all that concerned) a manky and ruined world then we should make sure that this project becomes a complete success.

But back to the latitudes. Just play around a bit with PVWatts and have a look how much irradiation comes down every month in different places. The web site which publishes PVWatts assumes that you work with your panels facing the equator and an inclination suitable to the location (near to optimum). These are the standard setting; leave them as is don't worry about other inclination angles now.

PVWatts have data sets for many places in the world and when you look at the data for places near to the equator you will see that the radiation per month stays pretty stable over the year. The further you go away from the equator the higher are the deviations from average over the months.

I will now evaluate this observation from an engineer's point of view. If I can rely on the fact that there will be a steady irradiation every day I can make a design where I get the optimum amount of electricity at a minimum cost (days with less irradiation will be covered shortly). In such a situation I know how much energy will be needed, I can calculate the losses and then I know how many solar panels I will need. It's not that simple in each and every case but it can be done.

When I know how strong the deviations of the values are over the year then I can make my plans in a way that there will be minimal investment costs and a maximum kWh produced - that, after all, is the bottom line.

In order to make the idea absolutely clear I'm going to compare the solar irradiation (using data from PVWatts) for three locations. The numbers are kWh per day per square meter with good orientation.

8.1 A few examples of irradiation

First we have **Catacamas in Honduras**, 14.9° North

Jan.	Feb.	Mar.	April	May	June	July	Aug.	Sep.	Oct.	Nov.	Dec.	Avr.
4.60	5.43	5.90	5.88	5.43	4.74	4.72	5.16	5.29	5.11	4.62	4.49	5.11

Between December (month with lowest irradiation) and the average we have a deviation of 13%. Between March (month with highest irradiation) and the average we have a deviation of about 15%. For a technical system an engineer would say that these numbers are good enough.

Now we look at **Amarillo/Texas**, 35° North.

Jan.	Feb.	Mar.	April	May	June	July	Aug.	Sep.	Oct.	Nov.	Dec.	Avr.
4.71	5.54	6.14	6.45	6.23	6.52	6.55	6.25	6.07	6.37	4.74	4.75	5.86

Between January (lowest irradiation) and the average we have a deviation of 20%. In July this difference is roughly 11%. In total the irradiation is higher (compared to Catacamas) but the differences over the year are higher as well. It is slightly more complicated to make a good design of the power plant but it could be done.

Last but not least **Hamburg (Germany)** 53° North.

Jan.	Feb.	Mar.	April	May	June	July	Aug.	Sep.	Oct.	Nov.	Dec.	Avr.
0.81	1.80	2.94	4.24	4.77	4.43	4.15	4.10	3.09	2.06	1.26	0.63	2.86

Between the irradiation in December (lowest irradiation) and the average we have a deviation of 350%! In May this deviation is a mere 65%! The deviations are really large and the measures to compensate them will be problematic. A technician would immediately ask whether it is really necessary to build a solar power plant in this particular spot. In other words: if you insist on building a solar power system in Hamburg the kWh will cost much more than in Amarillo, not only because the irradiation is on average only 50% of the irradiation in Amarillo but also because the deviations over the year are so much larger.

If you make the design (of your PV installation in Hamburg) in such a way that you have enough energy in December then the power plant is too big in May by the factor of 7.6 i.e. about 85% of the plant is not used (we wanted to cover just our own consumption and not feed into the grid). Most of the time the biggest part of the installation is just standing around (and costing money) because it will be needed in December. In this scenario we do not even allow for long periods of bad weather - two weeks with rainy days are not all that uncommon in northern Germany. The reality is gruesome enough even without rain. But this result should not come really as a surprise to anybody who is seriously thinking that planting pineapples in Alaska is a good idea (or photovoltaics in Hamburg).

We have to modify the statement from the start of this book that feeding into the mains is a stupid thing to do (from a purely economic point of view). Commercial solar power plants in Hamburg would be economically senseless if the possibility of feeding into the mains during the month with more irradiation did not exist (this statement was made in 2013; if the prices of photovoltaic panels keep going down at the same rate as they did during the last few years, I might have to modify this statement again in the near future).

So we do not only have large variations of the amount of irradiation day by day but also large variations over the year. There must be additional power plants which can store energy (i.e. fuel of any kind) and which can adapt relatively fast to these variations (big nuclear or coal power plants can't). From the capitalist point of view that will be perfect for the owners of such power plants: they get a guaranteed profit! Whether the power plants work or not, the owner gets paid. This will be a real necessity in Germany (or Alaska) but in large parts of the world it would be dispensable luxury.

8.2 The 37°-belt

It is not a fixed rule as such but there is some kind of consensus under technicians that says that a system should be able to ignore variations of about 10% (plus or minus) and should be able to cope with variations of 20% (plus or minus). Larger variations than these will almost invariably cause expensive problems!

Using this rule of thumb we more or less have the definition of where in the world photovoltaics can be used without big problems. Ultimately the number is arbitrary so I took the freedom to set it at 37° North and 37° South (I wanted to include at least a small part of Spain because quite a lot of foreigners live there who now might want to buy this book). Up to this limit you can obtain full advantage from photovoltaics. That does not mean that outside of this 37°N to 37°S belt solar technology is useless. It only means that you have to calculate a little bit more.

A rock bottom simplification is that anybody with enough money to buy an old car or a new scooter has enough money to be able to afford a solar installation or to participate in one. When looking at the 37°N to 37°S belt we see that roughly 3 billion people are living there with enough money: parts of the USA, all Middle and South American states down to Brazil, India completely, big parts of China, Australia, South Korea, all Japan south of Tokyo, Taiwan, Thailand, Malaysia, all Arab countries and South Africa. If we subtract those who at the moment can't even effort a bicycle, that still leaves us with about one billion people who can effort electricity from solar panels and who would directly benefit from cheap electric energy during the day.

This benefit will create wealth and more spending power (without harming the environment) and more people will be able to enjoy cheap energy. Nearly all the countries just mentioned have to import fuel. Year in, year out. There are of course various calculations about how long it takes to generate the energy which has to be spent on the production of a solar panel. This time is generally less than 2 years and in the 37°-belt you can expect this time to be just one year (and the biggest part of the price of panels are the energy costs). What do you think is going to happen if it becomes possible to produce silicon wafers with solar energy within the 37°-belt?

8.3 Enlarged zone: the 55°-belt

Near to the equator there are no seasons. Over the year the irradiation varies only because of clouds. Further away from the equator the influence of the seasons gets stronger. In Amarillo (Texas) the ratio between the month with the strongest irradiation and the one with the lowest irradiation is 1 : 1.4 and in Hamburg this ratio is 1 : 7.6. If an installation in Hamburg is designed for full coverage in December then in the period May to July there will be a big over-capacity, which is not cheap.

In Amarillo we have an average irradiation of 5.86 kWh per day per square meter. In Hamburg it is 2.86 kWh. Over the year the kWh in Hamburg for direct consumption would cost about 5.2 cents which is still pretty interesting. If an installation is designed in a way to provide full coverage in December, each kWh in December would cost nearly 40 cents (5.2 * 7.6). This design does not make too much sense from an economic point of view.

Or we can calculate the other way round. We design the solar installation in such a way that our own consumption (on average) is covered in summer time. The energy not covered by our installation is bought on the free market which is perfectly possible in Germany (by law the suppliers of electricity are obliged to deliver electricity to any house or flat in any area which is designated officially as habitable). In this way we can increase the number of people who might benefit from photovoltaics; now we can include all the inhabitants of the rest of the USA (except Alaska), the south of Canada, all Europe (except Norway, Sweden and Finland and Iceland) and all countries in the southern hemisphere. Even the southern parts of Siberia will be included. That is, we now have at least one billion people more as potential clients for solar equipment!

If all of them could be supplied with cheap and emission free electricity, a big step would be taken in the rescue of the worlds climate. In addition, it would not even be necessary to spend money on this project, it would pay for itself! The only thing necessary would be to spread the word that it can be done and that it makes economic sense (for both people and countries). This knowledge has to be provided in such a way that as many people as possible can understand what has to be done and how it has to be done in order to save big amounts of money and the climate. That's why I wrote this book.

Just a short remark about Hamburg: when the topic was the EEG, I said that somebody who fed electricity into the mains was getting (summer 2013 in Germany) a Feed In Tariff of 14 cents per kWh. From this we derived the information that electricity production must be cheaper than 10 Cents per kWh, else it would not be profitable. When comparing Hamburg with Amarillo we saw that it might be possible to generate electricity for 5.2 cents per kWh, so that if we add the fee for a technical inspection and insurance we might end up at 7 cent per kWh.

If you feed electricity into the mains you earn 7 cents (minus income tax); if you use it yourself you save 20 cents! If the Energiewende was some kind of pork-barrel politics for the owners of large roofs, I could understand the whole thing. If it was not, then it must have been the biggest experiment in hypnosis ever performed: „Listen to me and do as I tell you: when you make economic decisions, you will prefer the option where you pay most taxes! You will love to pay taxes! Paying lots of fees will make you feel happy!" Human behavior sometimes causes really strange effects. Just in order to tap subsidies people throw their money away! In order to get money from the state one abandons the possibility of saving even more!

You see, our point of view has flipped. Although at one time it did not really make economic sense to produce electricity using just photovoltaics, it did begin to make sense if you had a second and reliable source of electric energy. In which case it makes sense nearly all over the world. If the sun is shining, you just use your own electricity. When the sun stops shining you simply switch over to the local distributor. Absolutely legal (until now)! The only problem is that such switches, if automatic, are either very expensive (a few hundred Euros) or they will never receive an official certificate.

I did a rough calculation for such an automatic switch capable of switching a load of about one kW and it should be possible to produce it (in a form that it could get all the necessary certificates) for less than 40 Euros. With CE-label, TÜV-stamp or GS-label the manufacturer would have a huge market (when I have finished this book I will sit down and design one). If it is not necessary to worry about official stamps and labels such a switch could be produced for about 10 Euros (strictly speaking you can use them within the EU as well but in case of any problems you have the full liability).

Apart from Spain there is not a single country in the world where it is illegal to set up a solar installation for private purposes (although the actual method used to make the set-up might be subject to tons of regulations). Normally there will be no problems as long as the panels do not change the shape of the building. In some cases the administration (or the town hall's beauty commission) will request that the panels can't be seen from the street or from some tourist attraction. So, try to find out about your local regulations and check which regulations can be ignored or are exempt from punishment.

8.4 Case studies: two examples from the 55° belt

If you invest some money in a basic photovoltaic installation you can drastically reduce your electricity bill, year after year, for the next twenty years or so. Let's see how that lovely trick is done (the basic irradiation data were taken from PVWatts).

We take a house in Düsseldorf in Germany, a typical terrace house from the 1950s, with a south facing roof. This part of the world is not known for its sunny climate and I selected Düsseldorf (51° north, 6° east) because I want to show what is possible. Instead of Düsseldorf I could have selected any town in Holland, Belgium, northern France or southern England.

On the roof, flush with the roofing tiles, Thomas Schmidt installed 25 solar panels of 100 Watts peak (Wp) each. Each panel measures 120 cm x 54 cm, which equals 0,65 m², and each panels has an efficiency of 15%; in total he bought 16.25 m² panels. Each panel cost €86.00 including sales taxes and delivery. All in all, Thomas Schmidt spent €2,150 on his panels.

He also has had to spend some money on the solar controller. This is a device that makes it possible to either feed his household appliances via the inverter or to charge the battery (or both). In addition he needs an inverter; this device converts the electricity coming from the battery to a type suitable for the domestic appliances (this is 230 Volts AC in Germany). Or he could buy a combined controller/inverter; one of these marvels is capable of delivering a maximum of 3 kW (enough to feed any one appliance in the household but not enough to feed them all at the same time; most probably the vacuum cleaner and the washing machine can run at the same time) and costs about €600.

Thomas Schmidt did not feel like monkeying around on the roof, so he asked a small building company to fix the panels on the roof for him. That cost him €500. This then is his investment:

25 panels of 100 Watt each at €86 each	2,150 €
3 kW combined controller / inverter	600 €
work on the roof plus materials	500 €
Total	3,250 €

Before Thomas Schmidt ordered the parts for his solar installation he had a very close look at his electricity bills over the last few years. His household consumed 3,280 kWh per year, which comes to 9 kWh per day. Then he added up all that he had to pay; this amount includes (this could be somewhat different from company to company or in different countries):

> a) a basic charge for the connection; this has to be paid whether there was any consumption or none

b) the number of kWh the household used
  c) and of course taxes.

Mr. Schmidt consumed 3,280 kWh per year and he had to pay €886. That comes to 27 Euro cent per kWh, as you might remember from the chart a few pages back (that makes Mr. Schmidt very average although he does not like being told that he is average).

Now we have a look at the irradiation data to be expected on average over the year (PVWatts automatically assumes panels facing the equator with near to optimum inclination; at this point in our story we don't have to consider this topic in detail as yet).

In the following table we have, in the first row, the average irradiation per month as given by PVWatts.

In the second row we have P_{out}, the power output of the 25 panels measuring 16,25 m² and with 15% efficiency. This power goes into the controller, which has an efficiency of 94%.

The third row shows S_{out}, the power coming out of the controller.

We assume that we are using an inverter with an efficiency of 90%. The amount of energy coming out of the inverter is shown in row number 4.

The last row shows the amount of energy Mr. Schmidt has to buy per day because he calculated the size of his installation in such a way that in summer time he could be self contained.

	Jan.	Feb.	Mar.	April	May	June	July	Aug.	Sep.	Oct.	Nov.	Dec.
kW/m²	0.82	1.99	2.49	4.16	4.46	3.63	4.22	4.36	3.08	2.43	0.92	0.80
P_{out}	1.99	4.85	6.07	10.14	10.87	8.85	10.28	10.62	7.50	5.92	2.24	1.95
S_{out}	1.87	4.56	5.70	9.53	10.21	8.32	9.66	9.98	7.05	5.56	2.10	1.83
W_{out}	1.68	4.10	5.13	8.58	9.19	7.49	8.69	8.98	6.34	5.00	1.89	1.65
Diff	7.32	4.90	3.87	0.42	-	1.51	0.31	0.02	2.66	4.00	7.11	7.35

We can see that Thomas Schmidt did a pretty good job with his calculations. In May his household runs completely on alternative energy. At the beginning of his calculations he was thinking about feeding his surplus electricity into the mains but decided not to (he hates all this bureaucratic paper shuffling) and he simply does not care that in May he has to waste 0.19 kWh per day. And he is really happy about the fact that in the year's average he only has to buy 3.29 kWh per day.

He finances the €3,250 for his solar installation with a loan over 20 years at an interest rate of 5% (since Mr. Schmidt has the house as a security these conditions are completely realistic; these will be the standard numbers used in calculations all through the book).

Back payment per day	44.52 cents
Interest rate per day	22.26 cents
Electricity from the grid is 3.29 kWh * 0.27	88.83 cents
Daily payment	155.61 cents

With a consumption of 9 kWh per day we come to a price of 17.29 cents per consumed kWh. That is roughly 10 cents cheaper than the actual price of electricity. Thomas Schmidt can cut down his electricity bill by 36% (that he will need a battery as well I left out for the moment)!

Juan Portales is the owner of a house near Palma de Majorca in Spain. The house next door is owned by the Schmidts from Düsseldorf. Herr Schmidt had been talking enthusiastically about his solar installation at home in Germany. Could that be something for Juan Portales? Well, there is much more sunshine in Spain but on the other hand, one only pays 22 cents per kWh which is about 20% cheaper than in Germany.

Juan Portales has a look at his electricity bills over the past year. He adds it all up and divides the sum by the number of consumed kWh. For 3,285 kWh he paid €722.70. The result works out that he paid about 22 cents per kWh - which is slightly above the European average. Juan Portales would like to cut down his costs as well.

First he has a look at the calculations of Thomas Schmidt, starts up his computer and visits the PVWatts page in Internet. The most important information for him is the irradiation to be expected in Palma during summer. PVWatts states that he can expect 6.72 kWh per day in July. Thomas Schmidt had an irradiation of 4.46 kWh in May and was then able to supply all the electricity he needed using his photovoltaic installation. That means that Juan could do the same in July using just 17 panels.

```
17 Panels of 100 Watt each at €86 each        1,462 €
Combined Controller / Inverter                  600 €
Work on the roof plus material                  500 €
-----------------------------------------------------
Total                                         2,562 €
```

Juan does the same calculation that Thomas Schmidt did but using the irradiation data of Palma de Majorca and just 10.4 m² of solar panels with 15% efficiency:

	Jan.	Feb.	Mar.	Apr.	May	June	July	Aug.	Sep.	Oct.	Nov.	Dec.
kW/m²	3.73	4.53	5.19	5.81	5.89	6.35	6.72	6.54	5.82	4.55	3.21	3.30
P_{out}	5.82	7.06	8.09	9.06	9.18	9.90	10.48	10.20	9.07	7.09	5.00	5.15
S_{out}	5.47	6.64	7.60	8.52	8.62	9.31	9.85	9.59	8.53	6.66	4.70	4.84
W_{out}	4.92	5.98	6.84	7.67	7.76	8.38	8.86	8.63	7.68	5.99	4.23	4.36
Diff	4.08	3.02	2.16	1.33	1.24	0.62	0.14	0.37	1.32	3.01	4.77	4.64

Again we finance the installation with a loan over 20 years at an interest rate of 5%.

```
Back payment per day                            35.09 cents
Interest rate per day                           17.54 cents
Electricity from the grid is 3.29 kWh * 0.27    48.84 cents
-----------------------------------------------------------
Daily payment                                  101.47 cents
```

Since he and his family consume the same 9 kWh per day that the Schmidt family used, the effective price per kWh consumed goes down to 11.27 cents. That is roughly 11 Cent less than what he has to pay at the moment. Juan could cut down his electricity bill by 50%, if it wasn't for this government in Madrid.

Juan starts to grin. Even if he has to pay the 6.2 cents per kWh his government is asking, then his electricity would still be 5 cents cheaper (not saving 50% but merely 20%, which is not bad either)! And if they go and increase this special tax once again, he would invest some money in big batteries, buy a small generator and cancel his contract with the electricity supplier. What else he said about the Spanish government I had better not translate; children might get hold of this book.

Then Juan becomes very quiet and contemplative. He says: „Thomas, just now the calculation is based on the assumption that our consumption during the summer is only just covered. On average I will buy about 2 kWh per day from the electricity company and I pay 22 cents per kWh for any electricity I buy from them. On the other hand I produce 7 kWh every day and those kWh just cost me 7.52 cents each. What happens if I just throw away some more extra energy? Then it all should be even cheaper for me. Will it?"

After a short discussion they agree that the only way to find an optimal solution for this problem would be to use a spreadsheet program. Then Juan says that he would try out and do the calculation using the same number of panels that Thomas has in Germany. On the part of his roof facing south there would easily be enough space for 25 panels.

	Jan.	Feb.	Mar.	Apr.	May	June	July	Aug.	Sep.	Oct.	Nov.	Dec.
kW/m²	3.73	4.53	5.19	5.81	5.89	6.35	6.72	6.54	5.82	4.55	3.21	3.30
Pout	9.09	11.04	12.65	14.16	14.35	15.48	16.38	15.94	14.19	11.09	7.82	8.04
Sout	8.54	10.37	11.89	13.31	13.49	14.55	15.39	14.98	13.33	10.42	7.35	7.56
Wout	7.69	9.33	10.70	11.98	12.14	13.09	13.85	13.48	11.99	9.38	6.61	6.80
Diff	1.31	-	-	-	-	-	-	-	-	-	2.39	2.20

25 panels of 100 Watt each at €86 each	2,150 €
Combined controller / inverter	600 €
Work on the roof plus material	500 €
Total	3,250 €

We come to the daily costs.

Back payment per day	44.52 cents
Interest rate per day	22.26 cents
Grid electricity charges per day	0.11 cents
Daily payment	66.89 cents

With his consumption still at 9 kWh per day Juan now pays a net price of 7.43 cents per kWh. Even if he gets forced by his government to pay 6.2 cents back-up tax, then he would get each kWh for 13.6 cents which is 8 cents below the regular price.

If he decided to save a few kWh in winter time (a family living a little bit energy-aware has no problem at all to live nicely with 5 kWh per day), he would have a self-sufficient energy supply. Ok, he would need some batteries and a small generator for emergency situations, but he could stop his contract with the electricity supplier entirely. Shot from the hip, the price for a kWh would come to something like 10 cents (can you vaguely remember this number from before?).

Juan's smile gets even broader. He looks around and says: „Over there is North, then Madrid must be somewhere over there." He sticks out his fist in that direction, back of the hand facing down with the middle finger sticking up to the sky. His face is a bit red but that could be due to the red wine or the hours at the beach in the afternoon. He seems to be very content with himself and decides that next day his son Pablo would have to show him how to use this spreadsheet program. Herr Schmidt has got pretty similar thoughts: „The more you throw away the cheaper it gets? Could it be that there is a major fault in politics?"

You don't need any kind of special training in order to set up a solar installation. It is much simpler than many companies would like us to believe. What I did in this chapter was to make their profit the profit of Thomas and Juan. Well, yes, there are some more stumbling blocks (I had to say this because else you would not read the complete book).

9. Are PV installations dangerous?

It is a myth that fire fighters let houses burn down completely because they dare not extinguish the fire because of the photovoltaic installation. At least in Germany, **not a single case** like that is known since photovoltaics were launched - and that was about 30 years ago.

An analysis of the statistics showed that during the last 20 years there have been 120 fires where the photovoltaic installation was the cause of the fire. Knowing that actually in Germany there are about 1.3 million photovoltaic installations, one can calculate that the risk is pretty low. Additionally there is no particularly high risk for firefighters to extinguish a fire in a photovoltaic installation (not even at distances between one and five meters) which is important, especially in the case of commercial installations (which use up to 1,000 Volts DC).

In spring 2013 the Fraunhofer-Institut for Solar Energy Systems ISE (the biggest German research institute for photovoltaic systems) together with the TÜV Rheinland (a very big company specializing in technical inspections) organized a workshop with 120 experts. Manufacturers, scientists, fire fighters and insurance people agreed that photovoltaic installations are different from other electric installations but no more dangerous as long as the installation complies with current regulations. The best fire protection would be if the installation was carried out by a fully qualified expert.

The main characteristics of photovoltaic installations are that they work with DC and that they can not be switched off completely. When light falls on the panels, they produce a tension. It has already been mentioned that a battery can be used directly for electric welding; with DC it is relatively easy to provoke a stable flashing arc (when using an AC current that is much more difficult since the electrons change their direction 100 times per second; at these moments of change there is no current and the arc collapses). Flashing arcs are the biggest danger in photovoltaic installations.

When somebody botched up some of the work or used low grade material in an installation, it could happen that a connection becomes loose and a flashing arc is generated which does not stop straight away. Such an arc is so hot that even metal melts, so that all the material around the arc will melt and catch fire if is combustible. Such melting increases the length of the arc; when the arc becomes too long (this depends on the voltage) the arc will self-extinguish. If the arc could set other material burning while active, one might have a major problem (please note that solar panels as such do not burn very well but they are inflammable).

If you have the solar panels on a wooden roof covered with tar paper, then there is the risk that the fire spreads fast and wide because the fire will provoke a stack effect between the roof and the panels. If the panels are on top of a metal roof then the defect cable will melt away without even heating up the metal below it all that much. If the panels are mounted on a roof with regular roof tiles, then such a fire has no chance of spreading.

9.1 How could a flashing arc start?

Ultimately only if the work is botched. (By botched work I mean work of poor quality that you can't notice directly!). So we have to look at where the work could be botched.

First we have the solar panel itself. The different cells within a panel are connected to each other by narrow strips of thin aluminum foil. If there is a faulty contact one could possibly see some sparks, but the problem solves itself since the thin strips work like fuses and break the electric circuit. Something similar can happen within the cell, in which case there are no problems either. The cell, or a part of the solar panel, simply stops working.

Cables laid properly (i.e. not swinging wildly in the wind and abrading as they rub against other components or materials) will not break. So the most probable cause of any problem is some or other connector, such as those that connect the cable to the plug or the plug to the socket. The plug-and-socket connector is designed so that it is nearly impossible to produce problems here. This leaves one trouble-maker: the connection between cable and plug or cable and socket.

In the next chapter „You want to give it a try" I describe how such a connection is made using crimping tongs. In crimping, the wire strands of the cable are introduced into the bush of the connector and the tongs press the bush deeply into the wires. The force is so high that wires and bush get connected by cold welding; this connection is gas tight and very reliable (which is why you can find this technique used in all cars).

Is the crimping not done with enough force then there will be some electric contact (at least at first) but the contact is not gas-tight. Therefore corrosion starts and the electrical resistance between the cable and the plug (or socket) will increase with time. Because of the higher resistance this area will get warmer and the warmer it is the better the conditions for corrosion. This is something that happens deep inside the connector and there is no way to see the defect directly.

Is it possible to detect these weak spots? Yes, there are some possibilities. The high-tech version is to use an infrared camera. Such a camera produces photographs in which the surface temperature is depicted in different colors and even small differences can be made visible. The disadvantage is that such a camera is obviously not cheap.

If you have to deal with a large PV-installation with different sectors and you have a separate solar controller for each section, then you can start to collect statistics. If any given sector starts to deliver less energy than the others (whereas previously the sector behaved normally) then you know the area where a problem sits.

Then there is the method of the poor man. Every few month you go with your fingers over the surfaces of the panels; you let them glide over every cable and you don't leave out a single connector. If it suddenly hurts because you touched something hot, you have detected a fault. If you can feel a spot which is warmer than the surrounding material, you have found a fault in the making. These faults in the making are like time bombs in which you can't see the dial.

These faults and 'faults in the making' must be isolated immediately. The affected cables or panels must not make any electrical connection to the installation any more, that is they must not be part of any circuit (if more than one part is effected you have to isolate all these sectors from each other as well). After this separation you can take your time for the actual repair but their immediate separation is necessary.

9.2 How to switch off segments safely

So, how to separate a PV-plant component safely? I start explaining that by first telling you how NOT to do it. Let's assume that you detected a connector which is much warmer than the others and the PV-plant is actually in use (that means that there is actually a relatively strong DC current). If you now disconnect the two parts of the connector then you might get carbonized finger tips! The reason is that we have a strong current and when pulling the connector out of the socket we immediately produce a flashing arc. Such an arc is hot enough to make the air conductive and to melt metal, and your fingers haven't got anything to oppose that arc.

As long as any current is flowing through your installation there is the risk of flashing arcs. So the first step is to switch off all consuming loads (that includes the inverter as well) and then pull out the fuse of the DC circuit (since you switched off all the consuming devices beforehand there is actually no current). If you have a circuit breaker switch in line with the fuse, you can use that instead.

No matter how high the voltage in the section, the circuit is open and the electrons can't flow. No flow of electrons means no flashing arc when now you open the connector (the voltage without a closed circuit can't produce a flashing arc).

9.3 Cable cross sections

It could well be that you are not allowed to work in your country with the data given in this chapter. If in any doubt, check it out. If in your country thinner cables are allowed, you had better stick to the sizes recommended in this chapter; in this way you stay on the safe side and thicker cables do not cost that much more.

Depending on where you lay the cables there are different maximum values for the current allowed through a cable. DIN 57100 differentiates between a cable that is laid with a distance to the next cable or if there is more than just one cable (and other stuff) in a tube. It might also make a difference whether you lay the cable in a house or on a boat and on a boat it makes a difference whether the cable is in a cabin or in the engine compartment. In the rest of this chapter I refer to standard copper cables with industry standard insulation.

The reason for the different cross sections is that cables heat up when current runs through them. Current means electrons moving and hitting the copper atoms. These start to swing and the kinetic energy of atoms and molecules is heat. After a while the cable might get so hot that the insulation gets damaged. Slowly the risk starts to develop that copper wires in different cables come into contact. That might provoke a short (which is not all that dangerous since we protect all the circuits with fuses) or a flashing arc (very dangerous since the fuse might act on the assumption that the current is still within limits).

The knowledge of how to size a cable comes from the experience of more than a hundred years and ensures that undamaged cables of any particular size will not get too hot. Whether AC or DC runs through a cable does not matter at all. The numbers in the following table are on the safe side and give the maximum current in Ampere that is tolerable in a cable with the given cross section (and I don't give any guarantee that the values are correct - check your local regulations!). The numbers, except for the first column, are in Ampere.

mm^2	DIN 57100	Germ. Lloyd	Car-Electrics
1.0	11	10	10
1.5	15	12	15
2.5	20	17	20
4.0	25	22	25
6.0	33	29	30
10.0	45	40	50
16.0	61	54	60

9.4 Energy losses in the cables

It has already been mentioned that a cable warms up when a current runs though it. We now need to have a closer look at this phenomenon. There are materials which are bad conductors, such as dry stone, glass, rubber or plastic. These materials are called insulators as well. Then there are materials that are good conductors, e.g. copper is a very good conductor. Lastly there are the so-called semi-conductors but we are not going to talk about them just now.

In Ohm's law we saw that, with a fixed voltage, the resistance determines how big the current is going to be. In order to be able to compare different materials with regard to their conductivity, the idea of specific resistance was introduced and there are long lists giving conductivity for all sorts of different materials. We, however, are only interested in copper since this is the material used for all standard cables (only for the sake of completeness: copper has a specific resistance of 0,0178 Ω * mm² / m; a single copper cable of one meter length and a cross section of one square millimeter has a resistance of 0,0178 Ω).

In order to make life easy I just set out this table with the necessary information:

mm²	Ω per meter (single direction)
1.0	0.0178
1.5	0.0118
2.5	0.0071
4.0	0.0044
6.0	0.0029
10.0	0.0018
16.0	0.0011

Lets assume that we have a house somewhere out in the sticks and our lighting system runs at 12 Volt DC. Our problem is that we can't drive the last 80 meters to the house because the only track is too narrow for our car. So we get the idea to have leading lights, one at the house and one at the car port 80 meters from the house. If, during the evening, we decide to go out, we switch on these lights and we can find our way to the car and its key hole; when we come back later that evening we don't need a torch for finding our way back to the door of the house.

We decide that a 10 Watt bulb will be sufficient (such a bulb as is used for lighting the car number plate). Installing the light at the house is no problem at all. We also have two 100 meter coils of 1.5 mm² cable; if we use these, will the light bulb at the car port work sufficiently, or work at all? After all, we need to use 80 meters cable there and 80 meters back again.

First we have a look at the lamp at the house. The cables here are very short so we can ignore their existence. The light bulb consumes 10 Watts at 12 Volts. We take the formula $P = U^2 / R$ and rearrange it in order to get the resistance:

$U^2 / P = R$ (U squared divided by P = R)
$= 12*12 / 10 = 14.4 \, \Omega$

The light bulb we want to install at the car port will have exactly the same resistance as the one at our house so we now pay some attention to the cable. We have to go 80 meters from the battery to the light bulb and then 80 meters back to the battery. The electric effect is the same as if we had a single cable 160 meters long and put the light bulb directly next to the battery. After a short glance at the table above we get a resistance of $160 * 0.0118 = 1.89 \, \Omega$ for the cable.

Now we come to something which is called a voltage divider. This is a little bit more complicated but still easy to understand. In the picture we have the resistance of the light bulb which we now call R_1 and the resistance of the cable now named R_2 (I did the drawing in this way so that it can be used for all similar problems). V_{tot} is the total voltage, so in our case it is the battery voltage of 12 Volts.

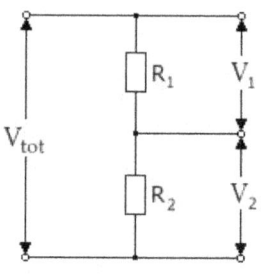

Now we have to use Ohm's law two times. **First** we have the voltage of the battery wanting to drive a current through the two resistors. As before, with the two parts of the cable, it does not matter at all, whether we have two resistors there or a single resistor with a resistance equal to the sum of the two resistors.

For a single resistor we would have:

$R_1 + R_2 = R_{ges}$ or
$14.4 \, \Omega + 1.89 \, \Omega = 16.29 \, \Omega$.

From $V = R * I$ (V equals R times I) which is the same as
$V / R = I$ (V divided by R equals I) or
$12 / 16.29 = 0.736$ Ampere,

which is the current that is actually going to flow through the wire and the bulb at the car port.

Now we come to the second utilization of Ohm's law. When a current flows through a resistor, it produces a voltage which can be measured over the resistor.

And we can calculate these voltages as well:

$V_1 = 14.4 * 0.74 = 10.66$ Volt

and

$V_2 = 1.89 * 0.74 = 1{,}40$ Volt

As a cross check we add the two voltages and - lo and behold! - we get our 12 Volts again (well, at least near enough), just as it ought to be.

And now we take our formula for the calculation of the electric power,

$P = V^2 / R$ (P equals V times V divided by R)
$V_1 * V_1 / R_1 = P_1$
$10.66 * 10.66 / 14.4 = 7.89$ W

and

$V_2 * V_2 / R_2 = P_2$
$1.40 * 1.40 / 1.89 = 1.04$ W

The sum of these two components is definitely not 10 Watts, as you might have thought. The light at the car port will be a bit more dreary than we expected. In fact, there is nothing we could do about that: the resistance of the cable steals part of the power and changes the voltage at the light bulb in a way we don't like. Love it or leave it!

The calculation of the leading light was some kind of academic exercise but now the principle should be clear. From now on we are going to make use of all this new knowledge in our photovoltaic installations. Our solar panels are on the roof and the electronics and the batteries are in the basement (neither batteries nor electronics like a hot or even warm environment; which is why they go to a cool place in the house, the perfect temperature being about 20°C).

I now have to anticipate the use of some facts which will be described a bit later in detail. Many solar panels, working at or near to their optimum, will deliver roughly 25 Volts (there are others with different voltages!). A family of two adults and two children with a demand that is a little above average will consume about 10 kWh every day. Most solar panels, where the cables are already connected to the panel, will be equipped with MC4 junctions. This type of connectors can be used for cables of different cross sections but smaller panels normally use cables with a cross section of 2.5 mm^2 so we use a cable with the same diameter so as to get the electricity from the roof into the basement. For the sake of simplicity the distance will be 10 meters from the solar panels to the solar controller (what a solar controller is will be explained in a moment).

We are going to neglect the fact that the solar controller, the batteries and the inverter for producing 230 Volt AC will incur losses. We assume 10 hours of sunshine per day; that means that our panels must deliver 1 kW so that in the course of the day we get 10 kWh. The question is now whether the decision to use cables with 2.5 mm^2 is sensible or not.

For comparing the cost effectiveness of different solutions we assume that the installation will last 10 years and the kWh we generate will cost 10 cents. I looked up typical prices for different cables (the prices are per meter for rolls of 100 meters; it is easy to see that the price goes up directly coupled with the amount of copper needed):

Cross section	Price in € per meter
1.0 mm^2	0.12
1.5 mm^2	0.17
2.5 mm^2	0.26
4.0 mm^2	0.42
6.0 mm^2	0.63
10.0 mm^2	1.05
16.0 mm^2	1.70

The panels deliver 1 kW and we are working with 25 Volts. Therefore the current must be 40 Amps (P = V * I or 25 * 40 = 1,000 Watt). We have a look in our table and see that a copper cable with cross-section of 2.5 mm² has a resistance of 0.0071 Ω per meter. The distance between roof and basement is 10 meters but we must go up again so that the circuit is closed. The cable has a resistance of 20 * 0.0071 = 0.142 Ω. With a current of 40 Amps flowing through the cable it creates a voltage of 5.68 between the ends of the cables. The formula P = V² / R gives the power loss with P = 5.68 * 5.68 / 0.142 = 227.2 Watt. The loss per day is therefore 2.3 kWh.

The cable costs us €5.20 and we have a loss by unwanted heat over 10 years of €839.50 (2.3 kW * 365 * 10 years * 10 cents).

We need to have a look at our losses if we used 4.0 mm² cable instead. Such a cable has a resistance of 0.0044 Ω per meter. The cable is again 20 meters long so we get a resistance of 0.088 Ω. Again we have a current of 40 Amps and it creates a voltage between the ends of the cable of 3.52 Volts. The formula P = V² / R gives us P = 3.52 * 3.52 / 0.088 = 140.8 Watt.

The cable costs us €8.40 and we have a loss by unwanted heat over 10 years of €513.92.

We need to have a look at our losses if we used 6.0 mm² cable. Such a cable has a resistance of 0.0029 Ω per meter. The cable is 20 meters long, so we get a resistance of 0.058 Ω. Again the current is 40 Amps and creates a voltage between the ends of the cable of 2.32 Volt. The formula P = V² / R gives us P = 2.32 * 2.32 / 0.058 = 92.8 Watt.

The cable costs us €12.60 and we have a loss through unwanted heat over 10 years of €338.72.

The main reason why I made all these calculations just now was to make sure that you wouldn't believe that the copper mining companies are behind my argument that thicker cables should be used. Additionally, I would be willing to bet that you did not notice that all these calculations were of no relevance anyhow (the calculations are all correct but they have no 'legal' basis). You can find the reason a few pages back. There is a table stating how thick a cable should minimally be for a given current.

We were making calculations using a current of 40 Amps, so according to the table the cable, 'legally' speaking, needs to have in minimum of 10 mm² cross-sectional area in such an installation. But if I told you that straight away you might have thought: „I'm living here in the middle of Bolivia! What do I care about German DIN-standards. I take the cables that I just happen to have at hand!" Now you know that this would be an expensive decision. Not in the immediate future but taken over the lifetime of the installation (and we'll ignore the possible cost of a fire caused by the cable being too thin).

So once again we have to do the same calculation and this time for the thinnest 'legal' cable. A copper cable with 10 mm² has a resistance of 0.0018 Ω per meter. The cable is 20 meters long, so that we have a resistance of 0.036 Ω. We have a current of 40 Ampere and that provokes a voltage between the ends of the cable of 1.44 Volt. The formula $P = V^2 / R$ gives us P = 1.44 * 1.44 / 0.036 = 57.6 Watt.

The cable costs us €21.00 and we have a loss by unwanted heat over 10 years of €210.24.

When we now compare the correct solution with what seemed to be the cheapest (and in most countries illegal) solution, we will spend €15,80 more for the cables. In compensation we will not transform electric energy worth €630 into heat. The thick cables are not there just to suit some legal principle, they really save you money. You might think about using even thicker cables.

For using thicker cables there are natural limits, the clamps of the solar controller. So, before you put this book down in order to go out and buy cables or dive into the Internet, have a look at these clamps (in case cable sizes are quoted as a diameter: circular area = $\Pi * r^2$). Additionally, you can assume that the manufacturers of such equipment know which cross sections can be stressed by which current. However, if you can make your own calculations it will give you the confidence that you set up a correctly working installation.

Now there is just one problem left to be solved. All solar panels have cables with a cross-section of 2.5 mm² but we must use a 10 mm² cable to go from the roof to the basement. Simple solution: we use a junction box. If you use a proper bushing, this box will even resist inundation. In the box we have a terminal block. From the thinner terminals we connect a cable to a single thick terminal. And on the other side of this terminal we connect the 10 mm² cable.

9.5 Connectors: MC3 & MC4 and Sunclix

MC stands for Multi Contact, an American company (owned by a Swiss parent company) that specializes in connectors for photovoltaic installations. These connectors have got several advantages and are actually a world wide standard; they are weather proof, UV-resistant and they allow an almost water-tight connection. Additionally (other than with human beings) only male connectors can couple to female connectors. This makes it nearly impossible to produce unwanted connections on the construction site (I'm going to talk about that in detail a bit later).

MC3 and MC4 connectors are available for different cable cross sections and you can buy connectors for 2.5 mm², 4 mm² to 6 mm² and 10 mm² cables (so there are three widely used sub-types) everywhere that you can get parts for solar equipment. MC3 and MC4 connectors are not directly interchangeable although there are adapters available (but it is much cheaper just to cut off the connectors and to set up a junction box).

MC3 connectors have the protection rating IP65; this means that they are absolutely dust tight (no possibility of touching conducting parts) and spray tight. There are three joints to be protected: cable to connector, connector to connector and connector to cable. For the protection of the transition between cable and connector a tight rubber tube is used. The connection between male and female connector is done using a gasket.

MC3 connectors were used in the past and they had two disadvantages. The first was that it was very difficult to join cable and connector without the use of special tools. The second disadvantage was that the two connectors were only kept in place by friction and suction (when the male entered the female connector, the air was pressed out; when trying to separate the two parts a low pressure volume developed that tried to keep the two parts together). Even with limited force the MC3 connectors could be separated.

The **replacement MC4** connector is different in two ways. The sealing between cable and connector is now made using a squeeze-type gasket - first the cap nut is pushed on to the cable then the gasket; then the connector itself is crimped to the cable and then the connector housing is pushed together with the other parts; and the last step is to tighten the cap nut in order to seal the gap between cable and housing.

In addition the two connectors are locked together; one of the two connectors has two hooks and when pressing the two connectors together these hooks latch into openings in the other connector. Merely trying to pull the two connectors apart does not separated the connection without actual destruction. The only problem is that when you try to disconnect them you need four hands: one hand grasps one of the connectors, the next two hands press the hooks inwards (for example using small screwdrivers) and hand number four tries to pull the other connector so as to pull the two connectors apart. If you got a pair of tongs capable of pressing the hooks together, the problem is reduced by one hand.

MC4 connectors are rated as being of protection class IP67; that means that they are absolutely dust tight (no way of touching conducting parts) and they are submersible.

The MC4 connectors are what should be used nowadays if the work on the roof has to be performed quickly (if you were thinking ahead, all the required cables could have been cut and assembled beforehand). Then each connection takes only a few seconds to complete. Oh, and the specification how many Amperes are allowed to run through an MC4 connector are not affecting the values of the table stating how many Amperes are allowed to run through a cable!

NB Use crimping tools for the connections. Else the connection might look right at first but will make problems later or might even start a fire! Never try to make a crimp using normal tongs; the connection will not be gas tight and corrosion will start increasing the resistance of the connection and making it possibly that hot that a fire starts! If you don't have crimping tools, ask in the car garage or ask an electrician to do the crimping for you. An even better solution is to order ready-to-use cables with connectors from Internet.

The only real competition to the MC connectors are the **Sunclix connectors** of the German company Phoenix Contact. These were the first connectors where you did not need any special tools (a small screwdriver is sufficient). These connectors come in two sizes. With the smaller version cables between 2,5 mm^2 and 6,0 mm^2 can be connected and the bigger one is meant for cables between 6,0 mm^2 and 16,0 mm^2. These connectors are as well rain and UV resistant and they come in two versions with IP65 (dust and spray resistant) or IP68 (dust resistant and submersible).

These connectors have as well a rubber gasket to secure the connection between male and female connectors. And they as well have two little hooks which prevent that unintendedly the connectors are separated. If you want to separate them by purpose then you take one side of the connection in one hand and introduce the little screwdriver into an opening (when you let go the screwdriver now it will stay in place) and with your other hand you just pull the connectors apart.

With the same simplicity the assembly or disassembly is done. Again we have a cap nut and the body of the connector (the squeezing type gasket is integrated) and we don't have a crimp-connection but a spring sheet. When this spring is released the cable can be inserted or removed with zero force. When there is a cable inserted and you press the spring into its position then the pressure secures a good electrical connection. Screw the cap nut tight (which presses the gasket together) and finished.

I think simpler and faster is not possible. As mostly when there is an advantage it comes combined with a disadvantage. These connectors cost a bit more. Of cause these connectors come with all the certificates you might ask for.

9.6 Fuses

First I have to correct a widespread misapprehension: fuses in a wiring system are cable protectors! They are not meant to protect any devices - they are only meant to protect the cables from heating up or even burning! Therefore you look for the thinnest cable in a circuit (cables with different cross sections in a single circuit must be avoided and if Germany were the only country to publish this book I would have to brand this as malpractice; no, I have to brand this as malpractice everywhere; done); with the cross section of this cable you go into the table and look up the permitted Amperes for this cable. This number gives the strongest fuse you are allowed to use in this circuit (there is no obligation to use such a big fuse; in the contrary, it is advisable to use smaller ones if at all possible).

In general one can say that in PV-installations the DC side is not dangerous, as long as the voltage between any two points is lower than 120 Volts, unless you provoke a short directly between the poles of the battery or you start a flashing arc. It might be that you have such an arc in your installation but the fuse still acts as if all values are in the "safe" area. Therefore you should add up all the possible currents of all the devices connected to this circuit and select a suitable fuse (but the value of the fuse must not be higher than the current allowed for the cable!).

We limit ourselves now to describing fuses of the type that have two metal caps with a thin wire stretched between them. The current running through the circuit runs through this wire as well and warms it up. If the current is far too high (for example in case of a short) then this wire runs through the following stages within milliseconds: rigid, fluid, gasiform, fluid and rigid again. But after the action is over it is wire no more but merely distributed material. The whole purpose is to separate the two metal caps electrically.

Some of these fuses are fast acting and others are 'extra fast' or 'time delayed'. Fast acting fuses need about 20 milliseconds to let the wire disappear. Medium fast fuses need 50 to 90 milliseconds for that and time delayed fuses take 100 to 300 milliseconds. However, in all of these cases the actual current was momentarily higher than the nominal current. Just remember: these fuses are there to protect the cable and not the devices! In fact, there are many devices which need a much higher current for a short period of time (nearly all electric motors and light bulbs). With the different types of fuses this behavior is tolerated up to different degrees. Let's suppose that the actual current is ten times higher than the nominal value of the fuse, but only for 2 milliseconds, then the cable is in no danger (if you replace a fuse with a solid piece of metal, you might find out that cables are nothing else but very long fuses, but you might just as well start up a nice camp-fire in your house and some of your devices might join in on the fun).

NB When installing batteries you should be extremely cautious because they can produce a very high current, heat up and start to boil, and then spill acid around. First step is to decide how strong a current is allowed to be going out or into the battery. Let's say we have again our 100 Ah Battery and it would really make sense to limit the current to 20 Amps (we are not going to start a car with it). Before connecting the battery with the rest of the installation you connect a fuse with the + pole of the battery and then you make sure that there is no possibility to come in contact with any metal between the battery and this fuse (you might use a plastic cap or some layer of silicone or some layers of sticky plastic insulation tape). On the other side of the fuse is the normal terminal. This little detail makes sure that, no matter how stupidly you behave, you will not be able to produce a short over the battery poles (saving you from a mess).

9.7 Safety regulations

DC voltages over 120 Volts and AC voltages over 50 Volts are definitively life threatening! All work on PV installation components which later might run at a higher voltage should be reserved exclusively for specially trained workers (you can drill holes into the walls and cut channels into them for the cables, you might even lay the cables, but it is not your job to connect them!).

Generally, even with a 12 Volt installation, one should follow the five rules for safe working with parts which are or will be (possibly) under electric tension:

1. Disconnect all poles: we make sure that there is not a single conductive connection between the part on which we want to work and any other part of the installation. When batteries are involved we disconnect both the plus pole and the minus pole.

2.	Safeguard against reconnection: perfect would be a padlock to make absolutely sure that nobody, by purpose or accidentally, can make a re-connection. This applies to all open connections. The absolute minimum requirement is a plate in front of the connection giving the name of the only person who has the right to reconnect.

3.	You check that there is no electric tension in the part on which you want to work (somebody might have installed an additional cable and you don't know about it yet).

4.	Earth and short the power lines of the area where you want to work. Since you checked in step 3 that there is no tension, this step can't do any harm. If some idiot ignored the safeguard described in step 2 or you forgot to secure a connection point, then only some fuses will blow but nobody will be in danger.

5.	Secure neighboring areas. With the steps 1 to 4 we made sure that we can work without any risk on one part of the installation. This step makes sure that you can't touch by chance another part of the installation which is not secured.

These measures might seem very formal at first glance; later on you might understand that this is exactly why they exist. After more than a century of handling electricity these rules have been developed and more than one dead body has been proof that these rules are essential.

There is one further rule: never work alone on an electric installation unless you must. If the other person understands (more or less) what you are doing this is a big advantage. If there is even the remotest chance that something (possibly) dangerous might happen, make sure that the other person knows how to shut down the installation completely. You do not want that lifesaver to become the victim of your poor planning as well.

Now once more so that it can sink into your brain: your fingers must not touch any part of an installation which, at a later point in time, could be subject to a voltage higher than 120 Volts DC or 50 Volts AC! Not even when wearing rubber gloves. It is not because you might endanger yourself, it is because you might endanger others!

10. You want to give it a try?

Are you one of those people who like to see something practical before they dive into the theory? Well, this chapter is just for you.

In a minimal system you need to have a solar panel, a solar controller, a battery and a consuming device (e.g. one or more lamps). Let's assume that you have an arbor in the garden and that it also would be nice if there were a light there and in the tool shed when you store everything away in the evening. Afterwards you can sit down in the arbor and read last year's newspaper in peace. Or maybe you are someone with a little house in the Andalusian hills and in the evening you need some light to prepare your dinner and read some poetry afterwards. Well, let's take a stroll through the Internet and do some shopping.

Just a tip for those of us who don't know the situation. If you live in the Andalusian sticks you might be able to use your mobile phone or even have an Internet connection, but you don't have a delivery address. So you need to find somebody who is willing to order the stuff for you, pay for you (for example using Paypal) and receive the goods when the postman comes. You think you can order that stuff yourself? Is your name there at the door bell? Sorry, but delivery is made only to the person who ordered the goods. We'll assume that you can solve this problem somehow.

The following prices were typical of those in the autumn of 2013 without specially looking for bargains, and with a delivery address in Germany. The size of the battery was selected so that the two LED lamps could give light for 7.5 hours (you can invite friends to your arbor and play cards all evening long or thoroughly read the complete newspaper that evening).

Solar panel, 20 Watt	27.95	delivery included
Solar controller, 10A	10.00	delivery included
lawn mower battery, 10Ah	26.80	delivery included
2 LED lamps, 600 Lumen	15.28	delivery included
2 switches + cable	10.00	to be bought locally
multimeter + small parts	40.00	to be bought locally
Investment	130.03	

You will need the multimeter (if you don't know what that is, have a short look at the supplement chapter about multimeters) and also a really sharp knife, a strip of screw (or luster) terminals and a few screwdrivers. If you don't have any of these, have a look at the local junk shop (the tools do not need to be of high quality now). We will also need some clamps, a fuse carrier and two or three fuses (have a look in a car spare parts shop). So we might end up having to spend 130 Euros.

First a word about DC. DC up to 120 Volts is not life threatening. We are going to work with 12 Volts; with this tension nothing dangerous can happen, unless you short the battery poles or you provoke a flashing arc. In order to make sure neither can happen you first connect the fuse carrier to the plus pole of the battery and then you make it impossible that anything can touch something metallic between the plus pole of the battery and the fuse carrier (plastic cap, a rubber or plastic sleeve, electrical insulation tape etc.). Now there is no danger in any part of the installation. The lamps will consume 16 Watts between them and the maximum charging current of the battery will be 1.66 Amps; a 2 Amp fuse will be fine.

Cables and flex can be bought by the meter; there is flex with one strand (there is also rigid wire as a single strand) and there are others with several strands. Cables with three strands are normally used for AC installations (230 Volts in most parts of Europe) and the different colored strands have the following meaning: earth is yellow-green striped (this strand must NEVER be used for any other purpose); Neutral should be brown and Phase should be blue. NEVER put your trust in the fact that an installation follows this agreement; it might be the last time that you were wrong. If you install a DC circuit these colors are taboo! This is for the very simple reason that somebody in the distant future might confuse a dangerous cable with a harmless cable. So, please, use cables with a single strand for you DC circuit and use RED for PLUS and BLACK for MINUS.

The argument „I know very well what cable I used for what!" does not count here. Its not so much that you might endanger yourself (a suicide attempt does not, as far as I know, carry a penalty anywhere in the world). The problem is that you might endanger others (for example if you sell the house and there are no complete plans about what cable goes where and why and how they are labeled). If there are AC circuits in the house as well as DC circuits, then the colors red and black are taboo from now on for AC circuits. If you have to call for an electrician for carry out some changes in the AC circuits, tell him explicitly what he must not do!

OK, just for a small shed and an arbor this was all a bit over the top, but I can't possibly know where and how you are going to make your solar installation. I don't want to scare you, and if you pick up my tip about putting a fuse at the plus pole of the battery there is no danger at all in our little installation. It is simply a variation on the old advice not to store dangerous liquids in bottles meant for drinks - you might kill your own grandchild. Therefore remember: take extra care with those things which could be dangerous!

Now we have a look at the solar controller, which in fact is a charge controller. Charge controllers in this price class have (nearly always) six terminals with screws. Two of them are for connecting to the solar panel, two are for connecting to the battery and the last two are for connecting to the consuming devices. So we have three groups and in any of them we have one terminal labeled '-' and the other one labeled with '+'.

We start with the easy part - the lamps. LED lamps for 12V DC have two connector pins. We cut off two screw terminals from a terminal block and screw one to each pin. Normally these lamps have a converter built in and it doesn't matter which way round they get connected (just to be absolutely sure you should have a look at the sockets of the lamps; if you find clear marks with '+' and '-' you should follow this hint). On the battery it shows which pole is positive (it's marked with a +) and which is negative (marked with a -). Now you take two cables (15 to 20 cm length will do fine), fit them to the lamp with the screw terminals and connect them to the battery (it is not dangerous to press the cable with your fingers against the poles). I assume the lamp will shine, so that bit is OK.

Now we make a big step. We fix the charge controller where we would eventually like it to be, put the cables and the switches for the lamps in place (you can use the standard 1.5 mm^2 cable normally used for electric domestic installations) and interconnect lamps and switches (you should use red cable all the way from the plus-terminal under the lamp symbol to the switch and from the switch to the lamp and black cable for the way back from lamp to the corresponding minus terminal; you will need the sharp knife for stripping off the plastic at the ends of the cables). Now you place the battery where you would like to keep it permanently (dry and cool if possible) and connect the battery with the battery terminals. Now a little LED on the charge controller will go on to signal that a battery has been detected. You have got light in the shed!

Now we come to the solar panel. If possible, the panel should be near to the charge controller and the battery. The orientation should be south to north (you don't need a compass for that; you wait until 12 o'clock noon and then you know how to set up the panel). The panel itself is a quite sturdy device but it presents quite a large surface for the wind to attack. Just putting the panel somewhere on the ground might be not such good idea. Fixing it with a rope is a bad idea as well, especially when the shade of the rope falls on the active part of the panel. Think of something to fix it firmly into place, but don't drill holes into the panel itself. Nowhere! Usually you will find pre-drilled holes in the aluminum frame of the panel. Use them.

Before you fix the panel you have to take care of the cables. At the back of the panel there will be a small box. Normally two cables will come out of this box, at the end of which are the so-called MC4 connectors. The cables and these connectors are suitable for outdoor use, that is they are rain resistant and the rays of the sun will not degrade them.

If these cables are long enough to reach to the charge controller the rest could be easy if the controller has got MC4 connectors as well, in which case you just plug them in. However, normally solar controllers do not have MC4-connectors, in which case we just cut the connectors off (you will loose the warranty for the panel by doing so; if you want to keep this warranty, proceed to the 'second case' a bit further on).

Now we need to use the multimeter. In order to be absolutely sure about which is + and which is -, we first check the voltage of the battery (the black cable of the multimeter is held to the '-' pole and the red cable is held to the '+' pole of the battery) and we see a positive number (if we get a negative number the cables are not correctly plugged into the multimeter).

Now we need to sort out the solar panel. We press the two probes of the multimeter into the copper of the two cables from the panel - one probe into one cable, the other probe into the other. If there is no minus sign on the display of the multimeter, the red probe is connected with the plus pole of the panel. It is a good idea to prepare small paper labels ready to fix to the cable with transparent sticky tape (if at a later date you have to change something, you will be happy to know which cable goes where and which polarity applies).

You connect the two cables from the solar panel to the appropriate terminal on the controller. Now another LED should go on, indicating that a panel has been detected and that the juice is coming in.

Done! You have just successfully launched your first solar installation!

The **second case** is that there are MC4 connectors at the back of the solar panel but the cables are not long enough to reach the charge controller. So we need to extend the cable. The easiest way to do that is to buy a MC4 cable from Internet (or a solar shop nearby). This cable needs to be twice as long as the distance between panel and controller; just cut it in the middle, identify which cable is plus and which is minus, strip the isolation from the ends, connect them with the controller and clip the connectors together. Done!

Case three is that there is no cable at all dangling from the panel. In this case you have to open the box and you will most probably find two terminals (if they are not labeled just use your multimeter). To make it possible to get cables watertight inside the box there will be two holes and each will be fitted with a washer and a cap nut with a hole for the cable. Get yourself sufficient 2.5 mm^2 cable and follow the instructions for the two cases above.

Every now and then you will hear that you have to use 'solar cable' for these connections. There are two differences between 'solar cables' and normal weatherproof cables. The solar cables have a second isolation so that they safeguard against voltages of 1,000 Volts and they are much more expensive. When you are buying the cables tell the seller what you want them for. Normal cables are meant for indoor installation and they are not weatherproof and UV resistant.

It does not really matter whether you connect an installation with 50 Watts or one with 50 kW. From the mechanical point of view the work will all be the same, but your planning needs to be a bit more extensive.

10.1 What did we gain?

In order to answer this question we first need to have a look at the panel. The one I happened to select at random was 555 * 355 * 23 (everything in millimeters). That is approximately 0.2 square meters (0.197 sq. m. rounded up).

It also depends on where in the world we live. I will just select some well known cities and we can take a look. In the first column we have the average irradiation per day per square meter of solar panel. In the next column we have the number of kWh coming out of the panel per day (assuming our panel is 0.2 square meters and reaches 15% efficiency). Now the energy has to pass the charge controller which has an efficiency of just 60% (we did not select a MPPT-type; we'll discuss later on what this means exactly). The battery also produces losses; I estimated its efficiency at 80%. In the last column we have the number of hours ONE lamp might shine per day on average so that the batteries could be fully charged once again the next day.

	Irrad. m^2	Panel out	Batt. in	Batt. out	Hours Light
Hamburg	2.86	0.086	0.052	0.042	5.3
Madrid	4.43	0.133	0.080	0.064	8.0
Bombay	5.01	0.150	0.090	0.072	9.0
Mombasa	5.43	0.163	0.098	0.078	9.8

We assume that the lifetime of the installation will be 7 years (which is equal to the time we write off the system). For better comparability I used an interest rate of 10% for all countries. The result is that the lighting costs 5.8 cents per day (4.3 cents back payment and 1.5 cents interest rate). In Hamburg the use of the reading light will cost a little over 1 cent per hour and in Kenya it will be about 0.5 cent per hour. Thus light from solar installations, so that the children can do their homework in the evening, is in Kenya affordable even for those considered to be poor there.

Just a remark why calculating the cost of solar equipment is so difficult. It is the same reason in Hamburg and Mombasa. It does not matter whether you use the available electric energy or not, you have the same costs because of the installation. If you think it would be a good idea to store the energy in batteries, then you pay extra for the mechanical abrasion inside the battery. This mechanism in the battery causes costs of about 10 Cent (or more) per kWh stored and subsequently used. If you find that hard to believe please read the chapter on batteries.

10.2 The Paraffin Lamp

When I was a child, it was standard for any household to have, in case of long electricity shortages, some candles and a paraffin lamp. The paraffin lamp is a pretty simple device. A small tank at the bottom, a wick to suck up the paraffin, a little wheel at the side to move the wick up and down (in order to adjust the brightness a bit). The flame was protected by a glass cylinder against draft and wind (in case you had to go outside).

Also there was the luxury version - the high pressure paraffin lamp. Again, this had a small tank at the bottom but the content could be put under pressure by the use of a little air pump. Paraffin flowed into the carburettor and was vaporized. The vapor was ignited and heated a gas mantle. This type of lamp was much more efficient and one liter in the tank was good for 8 hours.

This type of lamp was so successful that it is still used all over the world; they are often distributed under the Petromax name. In India you can buy them for about 975 Rupees which comes to roughly 15 Euros (in Germany they are four to ten times as expensive but then they get sold as camping equipment). We ourselves sometimes use these lamps for hobby purposes and occasionally as a light by which to work, but a quarter of the worlds population is totally dependent on them. Actually the world's population is about 7 billion people so that means that nearly 2 billion people are using paraffin lamps (Source: Wikipedia).

If each of them uses half a liter of paraffin per day then this comes to one million tons (just under) per day. Just to get a clear picture: these lamps use seven times as much petrol as runs through the tanks of all the cars and lorries in Germany! So we are talking about a considerable quantity, which does have an influence on the climate (and some people think that the damage caused to the health of people is even more serious).

It should be possible to do something about that using solar energy!

That at least was the thought of the German company Osram, who started a pilot project in 2008 in Kenya, as reported in an article on www.ingenieur.de (an offspring of the VDI-Nachrichten, the newspaper of the German Engineers Association) from 2009. The fishermen in Kenya go fishing at night and attract fish using paraffin lamps. For the paraffin they have to spend half of what they gain. So Osram lent them 11 Watt power-saving lamps together with batteries with a (useable) capacity of 0.1 kWh. The consumption of a paraffin lamp is about 30% higher than that of a normal light bulb and a power saving lamp only needs about 20% of a normal light bulb. That project seems reasonable.

Before that project was started the fishermen had to buy in average 1.5 liters of paraffin per day. One liter costs about 80 Euro cents so that they spent €1.20 every day on paraffin. Osram asked the fishermen for a 20 Euros deposit for the lamp plus battery and asked for one Euro for charging the battery the next day. Less pollution in the environment and the fishermen saved money as well. Looked like an exemplary project.

Since I have all the time a sharpened pencil by me and because it sounded too good, I made some calculations.

From the chapter about batteries we know that we have to calculate 10 cents for any kWh stored in a battery and used subsequently (you could say that this is the cost of wear and tear). We also know that a kWh generated in Kenya using photovoltaics will definitely cost less than 5 cents (including all costs as well as transport of equipment). The batteries have an available capacity of 0.1 kWh so using the battery costs one cent per charge to cover wear and tear and the generation of the energy will just cost 0.5 cents. So we have a total cost of 1.5 cents and they sell it for 100 cent! That is a profit margin that might make some drug baron's face go green with envy. But we don't want to be unreasonable - we have left out the labor costs.

The previously mentioned article states that the installation can charge 112 batteries within 3 to 4 hours, so we can assume that 200 batteries can be charged per day. If a local fisherman has a gross income of €2.40 per day (the €1.20 for the paraffin was half of his gross income), then the administrator of a charging station will, because he is such an important person, earn about 2 Euros per day and his helper and deputy (the man who does the actual work) will get €1.50 per day. So there will be labor costs of 2 cents per charging on top of the 1.5 cents we calculated earlier on. Now we have a cost of 3.5 cents and a selling price of 100 cents. That is still immoral and exploitation of the poorest for the profit of a multinational company!

Could it be done in a better way? Easily, no problem. Every fisherman willing to make a contract gets his own charging station. We begin by taking a 50 Watt solar panel. After 10 hours of sunshine we get 0.5 kWh. Since we only actually need 0.1 kWh for the battery itself we have 0.4 kWh reserve, which could possibly be used to charge a second battery for use by the family.

Such a panel you can buy on E-Bay for €57. Then we need a charge controller which costs a mere 10 Euros. The biggest cost will be the battery; batteries meant for lawn mowers (12 Volt, 20Ah, which equals 0,25 kWh) can be bought for 43 Euros (this size leaves a safety factor of 2.5). Finally, we need an LED lamp; they are a bit more efficient (about 20% less consumption than energy saving lamps) and much more robust when physically handled. What is more, they work directly on 12 Volts.

Solar panel, 50 Watt peak	57 €
Charge controller	10 €
Battery, 12 Volt 20Ah	43 €
LED lamp with 600 Lumen	15 €
Total	125 €

The situation in Kenya will be a bit different than the situation in northern Europe. So the fisherman will have to pay back his credit within a year and he will have to pay 20% interest rates. So we get:

daily payment interest	6.8 Cent
daily back payment	34.2 Cent
daily total	41.0 Cent

In the past the fisherman had to pay €1.20 per day for kerosene. Now he has a light for just one third of that price, starting right now. If he ploughs back his savings for just three months he will have enough money to buy (cash, not credit) a second battery and some small LED lamps. The nicest part of all this is that after one year the complete installation is his! For the next six years free lighting for house and boat and double his previous income! Over the full life time of the photovoltaic system, he will pay 5.8 cents per kWh.

That means doubling his standard of living within a year. That is real development aid!

10.3 Intermediate results

If you don't happen to know it off-hand, have a look at a map and find out on which latitude you are living and how relevant photovoltaics might be for you. I used the word "might" because there are a couple of boundary conditions.

Normally you can generate electric energy for less than 3 cents per kWh within the 37°N to 37°S belt. If you have a workshop somewhere in this region then you will need electricity for lighting and perhaps a few small machines. You couldn't care less that the installation does not generate electricity during the night. Or you have got some kind of office in which you work during the day and you want to use a PC plus printer and scanner. You are absolutely unconcerned about the fact that during the night there is no electricity in your office.

Maybe you are just asking yourself whether it might be possible to run a fridge or even an air conditioner with solar power. Yes, in principle you can, but there are some limitations to that so I will devote special chapters to discussing fridges and air conditioners. However, I would advise you not to jump directly to these chapters since first there is some more basic knowledge to be gained.

Electric energy can be produced within the 37°N to 37°S belt for less than 3 cents per kWh when the sun is actually shining. There is no other technology known today that is cheaper unless it is heavily subsidized.

All small businesses in the developing countries and the emerging nations could be powered with electricity generated by solar installations. For the owners of these small businesses it is of little or no interest whether there is electric energy available during the night; there would be no need for it anyhow!

How to obtain cheap light for the night was discussed in the last chapter - light for the whole evening costs a mere 6 cents. This means a drastic increase in the quality of life for billions of people (having a light in the w.c. protects people from being attacked by snakes not seen in the dark).

Yes, indeed, someone must be literate in order to obtain direct benefits by reading this book and even most of those who know to read will not understand the content when it gets a bit technical. However, simple solar installations are so easy to install that even academics with two left hands can do that - and academics can read. The only remaining question then is how big is their wish to help other people.

In Germany there has been much speculation and calculation on the question of how many jobs will be generated if alternative energy becomes really important. Introducing photovoltaics in the way described in this book would generate millions of new jobs in the so-called third world. Millions of people producing for themselves and their customers a considerable increase of the quality of life. Absolutely neutral to the environment and without requiring outside finance!

That brings me back to Kenya. This country, directly on the equator, has got hydro power but nowhere near enough. In Kenya, as in Spain and many other countries, most of the energy has to be imported in the form of gas and petrol, generating an external trade deficit. Any liter of paraffin not imported and then burned makes the country richer. Year, after year, after year. There is an alternative to the big fuel enterprises: just use the sun whenever possible!

11. The solar cell

Now, after more than 80 introductory pages, we arrive at the topic on which so many books have already been written. If you are interested in reading all of them then you can have (limited) reading fun over the next few years. Or you trust that I gathered everything that is really essential and even from all that you may only need a part.

The basic action of the solar cell uses the inner photoelectric effect. This involves a transition effect in a semi-conductor between differently seeded areas separated by an energetic barrier (what this actually means is not of much importance). When a photon hits an electron in the immediate neighborhood of the barrier, the electron can absorb the energy of the photon and jump to the other side of the barrier (yes, yes, it looks like some type of teleportation). This electron is now semi-stable on a higher energy level and it can be diverted so as to do some work. Afterwards the electron can flow back to the original side of the barrier (so we have a closed loop).

Pure metal is a much better conductor than the semi-conductor (as you might guess from the names) but the base material of the photovoltaic cell has to be a semi-conductor. The solution is to cover the underside of the cell completely with metal and to apply a metal structure like a honeycomb on the top of the cell (a complete film would not be a good idea since then no light could hit the energetic barrier in the semi-conductor). Part of the light is reflected or absorbed by the metal honeycomb but most electrons jumping through the barrier can be diverted. Here we have a technical compromise between the number of electrons set free (to work) and the number of electrons falling back behind the barrier without working first.

There are many semi-conducting materials and they all have been used (or at least tried out) in order to build photo-electric cells. There are two main parameters to all these cells. The first is the voltage they produce and the current they can deliver at a given irradiation level which is another way of stating their efficiency (how many photons are necessary in order to get a single electron to jump). The second parameter is the price, which can differ quite significantly as well. The material mostly used for solar cells is silicon because this is the most cost efficient (at least until now). Actually, solar panels can be made using cells from mono-crystalline silicon, from poly-crystalline silicon and from amorphous silicon. Also there are thin layer and thick layer cells. It begins to look as if the situation is a little out of control, everything starts to be complicated again!

So we make life a little simpler and scroll back to Juan Portales. He has a house with a roof area of 80 square meter (I forgot to mention that) and half of his roof is facing south; that means he can use, roughly speaking, 40 square meters. He only needs 16 m^2 for generating the electricity he actually wants. Therefore we need only buy those panels providing most kWh per actual Euro! Whether the panels need to take up 35% of his roof or 50% does not concern Juan Portales. But the fact that the solution that uses only 35% of the roof costs more than the system that needs 50% of the roof does concern him. So we just follow his pragmatic approach.

What counts are the Watts coming out of the system and how much they are going to cost us over the lifetime of the installation. Everything else is only interesting to those who want to show off. Yes, there are some people who need to make use of any surface, however small (boats, motor homes, studio flats etc.). Sorry, you people, I am not writing this book for you! In most areas where there is a lot of poverty there are also huge areas available for solar panels.

11.1 About light and shadow

Every now and then one hears, especially from the owners of yachts, that even a little bit of shade (like the shadow of a rope) is enough to cause the power output of a solar panel to fall drastically. So, lets have a closer look at what happens inside a solar cell.

We take a single cell (you can order them by mail for a few Euros) and connect our multimeter, switched to DC Volts. We cover the cell completely, for example with a book, and read a zero voltage; this is what we would expect. Now we lift up the book just a little and the voltage rises quickly to 0.5 Volts (if you get a much higher reading you most probably have not a single cell but a miniature panel). Now we take the book completely away - and the voltage stays at 0.5 Volts. Lastly we go out of the house and expose the cell to full sunlight. The reading stays unchanged at 0.5 Volt. The result of this experiment is that the voltage does not change (too much) with a change in the irradiation.

For the next experiment we cover our cell once more with a book and change the setting of the multimeter to DC Amps (we are going to measure the short circuit current). Now we push the book slowly to one side; the moment light reaches the cell there will be some current. If we now move the book back and forth over the cell, the current changes with the size of the area actually getting light. The area with light is directly proportional to the current.

Now we close the curtains of the room (thick curtains of course). The current will be pretty low. As we gradually open the curtains the current will gradually increase. We continue with our experiment and go outside into the full sunshine and the current will be much higher. The result is obvious: electric current is the actual number of electrons on the move; each photon is able to set an electron free (by seeing how many photons are needed on average to set an electron free we find out how efficient the solar cell is; if the efficiency is 15% then 85% of the photons are not doing their job). In practice, the number of electrons flowing through the cell is directly proportional to the number of photons hitting the cell.

Now we take a second solar cell and connect it in parallel with the first one; the current will double. However, when we cast a shadow on just one of the cells, the current will go down. If we cover one of the two cells completely the current is just half the value it was before. When we vary the amount of shade falling on to that one cell the current varies between 50% and 100%, as long as we don't cover any part of the other cell at all. That all seems to work as expected.

Next we disconnect the cells and connect them in line. The tension over the two cells has doubled. Again we are going to measure the short current and we find that the current has the same value as the current measured using just one cell. Now we slowly cover just one of the cells; the current get smaller and smaller and when just that **one** cell is completely covered there is no current at all. That is indeed a result that needs interpretation!

Now you should know that a solar cell is, electrically speaking, just a single electric component, a diode with a large surface. However, you could just as well regard the cell as being composed of a few thousand tiny diodes, all connected in parallel. From an electrical point of view these two ways of looking at a cell are absolutely identical. If you have a diode which is not hit by photons then this part of the area is not able to set free any electrons.

On the other hand, an electric circuit must be a closed loop. In a closed loop it does not matter where I measure the number of electrons passing by because at all places this number must be the same (if not, then some electron would mystically disappear in one place and reappear magically at another place). That means that if in all the cells connected in line the same number of electrons are on the move, then the cell receiving the fewest photons (making the electrons jump) is the one limiting this number.

A single solar cell delivers a tension of 0.5 Volts, which is not sufficient for most applications. That is why solar cells are connected in line; their voltages add up. There are solar modules on the market with different output voltages. Usually these are 18, 24 and 36 Volts but there are many others. By the way, solar cells (or modules) connected in line are called strings and modules connected in line got the same name.

Depending on the size you might find several strings (cells connected in line) in a panel with these strings connected in parallel in order to increase the maximum current. We now assume that we have a solar panel delivering 18 Volts with just one string. For many years yacht owners like to have solar panels for charging their batteries (at least partly) without any noise. Since often it is necessary to store these panels elsewhere, the owners prefer to use panels of a practical size (that means much smaller panels than those on the roof of a house).

In order to reach 18 Volts you need 36 solar cells. You can arrange them either as a panel of 6 by 6 (as a square) or 4 by 9, as a rectangle. With cells sized roughly 5 by 5 cm the rectangle would be slightly bigger than 45 by 20 cm; a perfect size for storage. Now you measure the power output of such a panel and then you move your finger so that its shade (about 2 by 8 cm) is cast on just one cell of the panel. Surprise, the current drops by 40%. Since the current must be the same in all the circuit the power output dropped by 40% as well.

Out of this situation the myth developed that just the small shade of a rope was enough to drastically lower the power output of solar panels. It was presumed that the power loss would increase further if more cells were effected. I think, if they had continued the experiment using a rope it would have caused a lot of head scratching.

But it is true, shade is a problem! If you have many panels (or strings) in parallel and any one of them can't deliver the nominal power that is perhaps a pity but not a great problem. However, if you have several panels connected in line in order to reach a higher voltage it can be a little catastrophic if a single cell is shaded. The whole installation does nothing but cost money.

There is a technical trick which does not actually solve this problem but does makes it bearable. The solution is to use what is called a free wheeling diode (if you can't understand the following explanation it does not matter too much). A diode is basically a component which lets a current run along in one direction but stops it completely going in the opposite direction. If such a diode is connected anti-parallel to a solar cell, there will be no current through this diode because the voltage has not the right polarity.

If we have a closed electric circuit with solar cells in line but one gets shaded, all the other cells would like to drive a current and build up some pressure. And this pressure is in the direction that this additional diode allows current to move. The solar panel as a whole must do without the voltage of the shaded cell (the output voltage goes down from 18 to 17.5 Volts) and over the diode in the direction which allows movement there will be a tension of 0.7 Volts (the tension lowers from 17.5 Volts to 16.8 Volts). With 16.8 Volts you can still do a lot and the nice thing about this method is that it works automatically.

However, it would be much better to set up your installation in such a way that no shade is ever cast on the panels. It may happen that it occurs in the morning and the evening, and you will have to learn to live with the effect of these shadows.

Some manufacturers push forward the fact that they included free-wheeling diodes in the panels as a unique feature of their panels, which is the reason why they are more expensive. If you are sure that you won't have any problems with shade cast on your panels then you can buy the cheaper panels without any loss.

11.2 Shorted solar panels. Dangerous?

Most people think that a short is something really bad (I assume that comes from viewing too many C-class movies). As an electronic engineer I can tell you that shorts can normally be avoided easily and they are definitely not evil or fatal - why do you think fuses were invented? Exactly, in order to prevent that something bad - evil or fatal - could happen. Unless you are working in the industrial energy sector, where they calculate using Kiloampere and Megawatts, a short is something completely unspectacular. Possibly you hear a short „Brk!" and it is all over. Thanks to the fuse!

With solar panels things are a bit different because we don't need a fuse for a single panel. That is because a short is one of the permitted conditions. A certain number of photons hitting the panel force an electron to jump. The more hits the more free electrons. If we now short the panel then the number of electrons jumping over the barrier will not increase; the current is self limited.

Let's say, before the panel got shorted, about 15% of the photons hitting the panel provoked a free electron; this energy, because it was consumed elsewhere, was not warming the panel. Now, with a short, this energy has to stay in the panel. What can it do? It can just warm the panel slightly. That is all that is going to happen when a panel gets shorted.

Something similar happens by the way when the opposite of a short occurs, namely the open circuit. The electrons can not flow so it is not possible to deliver energy to other places. The energy of the radiation has to stay in the panel and warm it up a bit.

If you think that this warming could possibly have a negative side effect, the answer is "No, it doesn't. The panels will not get warmer than a panel installed upside down."

11.3 Wp and the other cryptic information

Some decades ago solar panels were extremely expensive and were used only for satellites or spacecrafts. Later they were used in science when far away from any normal power supply data had to be gathered and batteries were no option. Up to this time solar panels (if you could call them that then) were one-off productions and they were tested individually.

About 20 years ago industrial production started (that was shortly after the German feed in tariff laws for alternative energy were first introduced). Many new manufacturers started their business and all of them had their own standards, some even for different lines of products within the same factory. However, since we Germans love to have everything properly aligned (that's so necessary if you want to sue somebody), we started to think about standards for solar panels; this standard then made it possible to compare panels from different sources.

That is how the STC (Standard Test Conditions) started. The test conditions had to be similar to real world conditions so it was seen as necessary that the light used had to be as similar to sun light as possible. That made it necessary to take into account that air is not completely transparent but filters some frequencies more than others. When you stand at the equator and you look at the sun, a 'standard air mass' exists between you and the sun. When instead you are in northern Europe you are looking slantwise through the air with the result that between you and the sun are 1.5 standard air masses. Since nobody else outside Germany was interested in a standard for solar panels, this value was incorporated into the standard. Now only a calibration point was missing, so they checked the irradiation data for Germany and found out that in Germany the irradiation level is hardly ever higher than 1,000 Watts per square meter. That was a nice round number and the standard was finished (well, even the Germans are not that fast and there is a bit more to these tests but I have to stay entertaining).

Wp is the abbreviation of Watt peak, so it is the maximum power output. For the test an artificial sun is used and the irradiation is so strong that 1,000 Watts of light reach any square meter of panel. Since in Germany you will hardly ever find conditions with a higher

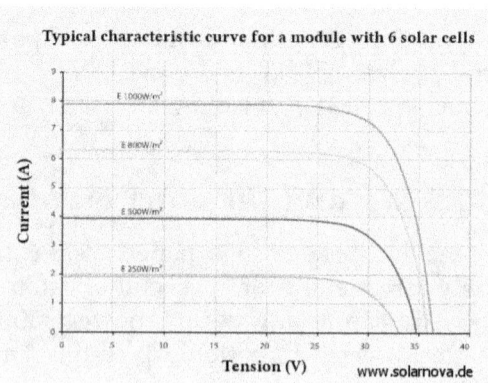

irradiation this is the maximum power input. More is possible, but not in Germany.

In the picture you see the characteristic curves of a solar module with an output voltage of about 24 Volts. The no-load voltage can be measured at the clamps when there is no current flowing. Depending on how strong the irradiation is this voltage is between 33 and 37 Volts (with this type of module). The no-load voltage is largely independent of the irradiation.

The next characteristic value is the short-circuit current. This current we find in the diagram at the voltage zero. It is easy to see that this current is directly proportional to the strength of the irradiation (this affirms our theoretical considerations).

Now we have to explain what all these curves mean. I take the curve for an irradiation level of 800W/m² and calculate some working points. You have to read the following table like this: if you attach a resistor of 0,79 Ω to the module the voltage will go down to 5 Volts and the resistor will consume 31.65 Watts (or better, it will convert 31.65 Watts electrical energy into heat). If the resistor has a value of 1.58 Ω then the voltage will go down to 10 Volt and the resistor will consume 63.29 Watts etc.

Tension	5 V	10 V	15 V	20 V	25 V	30 V	35 V
R = V / I	0.79 Ω	1.58 Ω	2.38 Ω	3.17 Ω	4.00 Ω	5.26 Ω	19.44 Ω
P = V² / R	31.65 W	63.29 W	94.54 W	126.18 W	156.25 W	171.10 W	63.01 W

The power which can be taken from the module increases (up to a certain degree) in nearly a linear fashion with the value of the resistor and then it drops drastically if the value gets any higher. This means that, even with full irradiation, a panel which is rated 100 Watts peak will normally not deliver 100 Watts. It could deliver 100 Watts if the resistance of the load were perfectly fitting to the panel. For the table I was using the 800 Watt line; why don't you, just for fun, calculate the nominal wattage of the module (I myself do know it but I'm not going to tell you just yet).

The working point on any of these lines where the maximum power is delivered is called MPP (maximum power point). I will tell in the chapter about solar controllers what happens to the MPP when the irradiation changes. Often the MPP is only given for the STC.

Then we have the module efficiency (this value is often not stated; instead only the cell efficiency is given; this value is generally higher than the module efficiency and might attract possible clients; but for calculating our installation we need to know the module efficiency). How to calculate that you will see in the next paragraph.

The cell efficiency is always stated (the standard requires that). This efficiency is always higher than the module efficiency because for this efficiency the spaces between the different cells and the frame are taken into account as well. By way of an example we take a look at the panels of Herr Schmidt. They have a surface of 0.65 m^2 and a nominal wattage of 100 Watts. For getting the module efficiency we need to find out how many square meters we need in order to produce 1,000 Watts (as required by the STC); with the panels of Herr Schmidt that would be 6.5 m^2. Then the module efficiency is simply P = (1 / 6.5) * 100 = 15.38%. In this way we can easily calculate the module efficiency ourselves and we don't get dazzled by the high cell efficiency parameter.

If you get irritated because you have to walk with your pocket calculator from shelf to shelf, needing a minute or two for each calculation of the panel's module efficiency, it might be a good idea to complain to the sales person or the manufacturer - maybe it helps in the long run. And don't accept the answer that the standard does not ask for that value. That is true but the standard does not say that you may not publish the module efficiency.

The last data I'm going to discuss is the maximum system voltage which often starts some confusion (in many articles it is claimed that PV installations are dangerous because there are high voltages of 1,000 Volts in them). Well, there are commercial installations where you can find 1,000 Volt DC over the strings (with such a high voltage the inverters for direct feeding into the grid can be designed a bit simpler which means cheaper). In privately owned PV installations one will hardly ever find voltages higher than 100 Volt DC. Mostly you will find in maximum 18 Volts in small installations (the why will be explained a bit later).

11.4 How to find the right panels?

This question can not be answered because normally with solar panels there is no right and wrong; we only can choose between more or less adequate. When you set up a new installation all the panels should be of the same make. This is for technical reasons but mostly it makes your work easier. If you are a semi-professional you might prefer bigger panels because you can install them faster (measured in square meters per day). If you do your first installation yourself you should use smaller panels and get yourself a helper and/or observer.

I must not and I will not say anything about different brands. There are black sheep in any family and in the production of solar panels there are Monday morning products as well. On the one hand there are long-established companies that have been producing solar panels over at least two decades. The mere fact that they still exist is proof that they deliver schlock only every so now and again. This provides a certain feeling of safety for which you pay some percent more.

On the other hand there are, for example, companies from China or Korea; hardly anybody knows their names (nor, often, how to pronounce them). These companies offer their products for prices which caused the EU authorities to saddle these products with an anti-dumping tariff (whether the anti-dumping allegation is really justified I could not find out). Normally you buy such panels via the Internet. So have a look at the customer reviews. This is not as good as an insurance certificate but might be very informative.

Generally there is no guarantee that any given company will still exist in three year's time. No matter where in the world it is located.

So I only got two tips worth mentioning:

1. try to find similar panels (similar in size and electrical parameters) that are comparable to ones from other manufacturers. If after a couple of years some of the panels become defect then there is a good chance that you can still find panels of the same size as a replacement, even if the original factory does not exist any more; you lose your original investment but at least you can repair your system.

2. there are a lot of forums in the Internet concerning photovoltaics. On http://www.photovoltaikforum.com they have a module database where more than 75,000 makes of solar panels are listed, sorted by manufacturer with the additional information how many members of the forum have actually used them. This might be a good indication as well.

The only problem with this site for you might be that all is in German, but the few words you really need to understand there you can look up in a dictionary or use Google translate.

11.5 Tracking-Systems

The sun radiates its energy evenly in all directions. If you want to catch some of this energy you need a surface area on which the photons can fall. You can concentrate them using special lenses or with concave mirrors (many sun power stations use a great number of independent small mirrors that as a whole build a big concave mirror). Or you take the photons just as they come, for example using a solar panel. The bigger the area of our installation seen by the sun the more energy you can catch.

Lets assume that we had a solar panel laying flat on the ground somewhere near to the equator. At sunrise the only thing that can be seen of our panel by the sun is just a horizontal line and we can't catch any energy at all (if we ignore diffuse irradiation). The higher the sun rises the more it can see of our solar panel. At noon the sun will be directly over the panel and the visible area is biggest. Then the sun sinks slowly in direction of the horizon in the West and the visible area gets smaller and smaller. Mathematically speaking the visible size of the panel is given by the sine of the angle between the eastern horizon and the sun (but that is not all that important now).

Now one might rotate the panel so that the sun can see its full size at all times. Near the equator the panel would need to swivel on just a single North-to-South axis, so that it faces East in the morning and slowly follows the sun to face West in the evening. We would need a motor to move the panel and some electronics so that the panel would track the sun perfectly. Such an installation is called a 1D tracker.

If we put up an installation away from the equator then it becomes impossible to align the panel perfectly if it moves on just the one axis; now we need two axes, one for the East-to-West movement and another one for the up-and-down movement. The installation is called a 2D tracker and will be more than twice as complicated. In addition, we have to take into account that the panel might have to resist strong winds with wind speeds of 150 km/h or even more. The installation must be able to cope with such wind speeds so that the panels don't become UFOs (Unwanted Flying Object).

The whole installation must be pretty sturdy and it needs a good foundation. On the site http://www.photon-international.com I found an article in English about a company producing such trackers. It said that just the tracker costs about 855 Euro without VAT per kW potential; with VAT we come to about 1,000 Euros. At the time of writing this book solar panels cost about 100 Euros per square meter and they deliver 100 Watts. For one kW we need roughly 10 square meters of solar panel and they cost about 1,000 Euros.

Now we come to the question whether such a tracker pays for itself. Using geometric calculations (we are not going into detail here) it was found that (depending on where in the world you are situated) an 1D-tracker can gain up to 30% more energy and a 2D-tracker even up to 45% (calculated over the year and compared to a fixed panel with a good orientation).

If we don't do any further changes to the rest of the installation we get in maximum 45% more energy (it could easily be much less) by spending roughly two times as much money.

Generally you could say that spending that much is only justified in some rare occasions, for example when it is impossible to place solar panels with a good orientation but if there is enough space to place a sturdy mast. If the price for solar panels does not increase by more than 100% then tracking systems are not a good option.

Additionally you have to take into account, that if you set more than just one mast, you need quite a lot of space between them so that the solar panels of one tracker can not cast shade on the panels of another tracker. So with trackers you will need quite a lot more space than with a rigid installation.

12. Solar controller

Before I delve further into this topic I first feel the need to make an observation: this is the part of photovoltaics where most lies are passed on by the sales persons. I haven't a clue whether they do it on purpose or because they don't know any better (maybe the technicians are too polite to correct them); sometimes I suspect that we find ourselves in the field of criminal activities. This forces me to deal with this topic in a much broader fashion than I had originally intended because quite a lot of technical terms need to be explained first, since they are often used in a misleading or even completely wrong way.

If I were just telling you what to buy, the only thing you could do would be to nod your head politely; my statements would not have more value than those of anyone else. That is why I would like to tell you what happens behind the stage curtains and explain the actual technical basis. This is because you could easily cut the output of your installation by 50% of its (potential) capacity by using a solar controller of the wrong type or you might even torture your batteries in such a way that they will die within a few months (whereas you expected them to last 7 years or even more).

The first charge controllers (that is in fact what solar controllers are) were copied (at least as far as the actual technology was concerned) from those used in ordinary saloon cars. At the very beginning smelly carbide lamps were used in cars when driving around at night. Then somebody had the bright idea of connecting a DC generator to the motor of the car and using electric light bulbs (what an improvement!). Since that was the only purpose of these generators, they were called "light machines" in Germany.

At that time the engine of a car had to be manually started using a crank handle. Then somebody had another bright idea; instead of cranking by hand one could use a small electric motor. The energy for this motor would come from a small battery and in order to charge this small battery one could use the light machine! Now starting the engine of a car was as simple as turning a key. The triumphal procession of the Motor Car really started because they became so convenient to use.

Batteries don't like voltages being too low and they don't like voltages being too high. Now there was the problem that the output voltage of the generator was directly coupled to the number of revolutions of the motor (we will come back to that). For quite a while this problem was ignored and the output of the light machine was directly connected to the poles of the battery. However, you had to make it clear to the driver that he should not drive long periods of time with very high revolutions of the engine in the day time or with very low revolutions when the lights were switched on. Since most drivers did not know about this, these early generators were real battery killers. Then somebody had the idea that it might be possible to regulate the output voltage of the generator. In order to be able to understand how that works we have to make a small excursion into physics.

We have a magnetic field (I explain in a moment how we generate it) and within this magnetic field we rotate a coil. If the magnetic field embraced by a coil is changed then a voltage is induced into the coil (physicists and electronic engineers have elaborate formulae for calculating these effects but they can't really explain what happens, so we just leave it at that). Seen from the view point of the coil the magnetic field comes once from the right hand side and a moment later from the left hand side. So what is induced is alternate voltage (better known as AC). With a given coil the tension of the AC voltage depends on how strongly the magnetic field is changed per time unit. In order to increase the output voltage of the coil we can either increase the revolutions or we can make the magnetic field stronger.

Since the generator is rigidly coupled to the engine of the vehicle (the revolutions cannot be easily or freely changed) the possibility left was to change the strength of the magnetic field. So now we are back to the question of how to generate the magnetic field. Well, we generate a magnetic field by simply using a coil. The strength of the magnetic field of a coil depends on the strength of the current running through it; that is, if you double the number of windings of the coil you double the strength of the magnetic field. When producing a static magnetic field with the coil then the current in the windings is determined by the voltage set to the coil and its resistance. In a coil with many windings a small change in the voltage causes a strong change of the magnetic field. Small changes of the input to the coil could bring considerable changes of the output voltage of the generator. Changes in the speed at which the generator turned could be compensated easily.

By losing a few Watts in the exciter coil (that is the correct name of the coil building up the magnetic field for the generator) one could control hundreds of Watts generated by the generator. At that time there were no electronic circuits available so a mechanical switch was used. This switch closed the circuit for a few milliseconds, the magnetic field of the exciter coil got stronger, after which the current was switched off again. If the switch-on time was longer then the output voltage of the generator became higher; if the switch-on time was shorter the output voltage went down (maybe you remember that the power grows with the square of the voltage; so with a small change of the exciter voltage we can control the power output).

For those who know what pulse width modulation means: this was the first time that it was used on a large scale (whether it was invented at that time, I don't know). For all others who don't know what PWM means: you will know in a moment. Whenever a DC current is switched on or off (and especially when it is switched off) there will be a spark, and sparks emit radio waves. If a car passed by using this method of regulating the battery voltage the people around only heard a loud "RRRRR" in their radios and saw nice white random pixels on the screen of their TV. Possibly the car drivers did not care too much about molesting other people (that is if they knew that they did so) but with the introduction of car radios they themselves were affected and this was the end of this type of regulator. From that time on electronic circuits (now generally available) were used and the sparks were not really necessary any more (for the retrieval of honor of the automobile companies I must admit, that most encapsulated the regulator well).

Now, knowing the historic background, I can explain to you how solar panels are connected to the batteries. We simply use the technologies developed for cars.

12.1 Direct connection

First I tell a little horror story. We simply connect the battery directly to the solar panel. We use an ordinary 12 volt battery with a capacity of 10Ah and connect it directly to a solar panel which can produce 100

Watt at 24 Volts. The perfect charging current for this battery (when completely empty) is one Ampere whereas the solar panel can deliver 4 Ampere in bright sunlight at noon. As far as the battery is concerned, 12 Volts (or a little bit more) would be fine but it in fact it gets force fed with 24 Volts. Since the battery can't disconnect itself, there are only two possibilities left of what to do with all that surplus energy:

1. The water in the battery liquid is split by electrolysis and the gas just disappears.
2. The energy is transformed into heat.

Without going into details (you can look them up in the supplementary chapter about batteries) the following will happen: the liquid level drops rapidly, then a short while after the battery will be (more or less) dry after which it will be dead. The warmer a battery gets the faster it will age; every 8°C over 25°C will halve its lifetime. If you are lucky you might not have to clear up a complete mess but your battery will definitely not survive longer than a few weeks (without maintenance).

During my investigations I found quite a few advertisements for this type of regulator, which stated that this technique was rock solid and had been successfully used for decades, but I never found a hint as to what it might do to your battery (the advanced ones at least disconnected the battery when full). If somebody tries to sell you this type of solar controller (under whatever fancy name they give their apparatus), steer clear of that shop in the future!

Now we take a second example. We connect a 5 Watt solar panel which can deliver a maximum of 14 Volts directly to a 100 Ah battery. The battery forces the output voltage of the panel down to 12 Volts and the panel will deliver a maximum of about 400 mA. The panel would need more than 250 hours of sunshine in order to charge the big battery fully. The self-discharge of the battery will hinder the attempt to charge it, and so we will need more than a month to charge the battery. A 5 Watt solar panel is no danger at all to such a big battery.

That means that you have to match the sizes of both the panel and the battery as well as take into account the irradiation data of the place where you have the installation (seeing as you will not be able to change the weather) in order to get a good technical solution. If you don't want to worry about the problem of how to get a perfect match then you should believe me that there are simple and fairly cheap methods for avoiding this problem. I'm going to show you them in a moment (the shunt regulator is no good solution either).

12.2 Shunt regulator

A shunt is a resistor which is connected in series with a consuming device (a consuming device is any apparatus that consumes electricity, such as a light or an electric motor). A shunt has a very low resistance and is normally used to measure the current in a circuit without influencing the circuit itself too much (the current going through the shunt provokes a voltage over the shunt; knowing the resistance of the shunt the current can now be calculated). But that is not what is done in a shunt regulator.

The classical shunt regulator is found in very old wind turbines, which are directly coupled to an electric generator. The wind turns the rotor of the generator and the energy is fed into a battery. When the battery is full the delivery of electricity to the battery must be stopped. If the electric connection was just switched off the wind turbine would have no resistance any more; it would therefore gather more and more speed and most probably be damaged eventually. So the connection is switched from the battery to a high power resistor and all the energy that had been flowing into the battery is now converted into heat. This energy is then lost to the surroundings but at least the turbine is not damaged, since it is turning against a resistance once again.

In the chapter on solar panels we saw that in a panel all states between fully open contact and completely shorted contact are permitted, and no harm is done to the panel regardless of how long any one state continues. When the battery is full it is not necessary to shunt the power; one simply does not use that surplus power. What actually happens if we allow this is something we have already seen: the solar panel just gets a little warmer, like cardboard of the same color. That's all.

If somebody tries to sell you a shunt regulator for your solar installation then either the sales person hasn't a clue (looking at the situation from a charitable point of view) or he is a cheat, trying to sell you some surplus components for a large amount of money! You should ignore immediately everything that is labeled with 'shunt' and 'solar' and never ever go into that shop again (if you found that offer in the Internet then tell your firewall to block that particular address from now on).

12.3 The linear regulator

This kind of regulator was state of the art about 40 years ago and it is still used when small power supplies are needed (just a few Watts at low voltages). If for example you have a 12 Volts power supply but need additionally 5 Volts then this is something you might search for in your junk box. If it needs to be a low consumption solution you would select a DC-DC converter (we will come back to these devices soon).

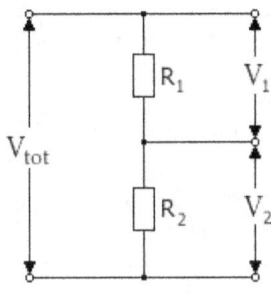

What a linear regulator does can be understood easily if we look at the voltage divider diagram which we already know. V_{tot} is the output voltage of the solar panel. R_2 is our battery and R_1 is our regulator (its resistance can be regulated over a large range). If the voltage over the battery (V_2) is too high then the resistance of R_1 needs to be higher; this increases the voltage drop over the regulator and the voltage over the battery goes down by the same amount.

If the input voltage is much higher then the desired output voltage then quite a lot of electric energy has to be converted into heat. If the difference of the voltages is small then this type of regulator is convenient for installations needing no more than 20 Watts.

I will just give a standard example. We have a solar panel with an output voltage of 18 Volts and we want to charge a battery at 12 Volts. That is, over the regulator we must have a voltage drop of 6 Volts and that means that the power going into the battery is twice as high as the power being converted into heat, so we have an efficiency of 66%. Since the working point of the solar panel will only by chance be at the maximum power point the efficiency will normally be much lower.

12.4 PWM regulator

In effect the linear regulator fritters away all the surplus energy not actually needed (it simply transforms it into heat) just so that the required output voltage is obtained. Now we can have a look whether the PWM regulator is perhaps a better technical solution.

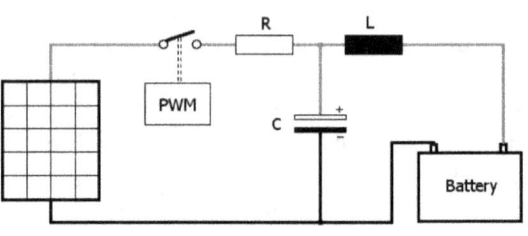

In order to be able to evaluate that, we need to understand how a PWM regulator works (by the way, PWM stands for pulse width modulation). We already came across this type of regulator when talking about cars and the generators used in cars. Now we have a closer look at this type of regulator; in the picture we have included all necessary components.

The solar panel must have an output voltage higher than the battery voltage. The capacitor was already charged from the battery via the inductor L. If now the switch of the PWM regulator is closed then there will be a current through the resistor R (it is only needed to delimit the current so that the switch of the regulator is protected) into the capacitor. Coils are conservative in the sense that they don't like changes; therefore the capacitor is charged (the voltage over it gets higher) but the current through the coil stays (more or less) constant. The switch of the regulator opens again and the voltage of the capacitor slowly increases the current through the coil. After a short while the voltage over the capacitor will be smaller than the voltage over the battery and the coil driven current into the battery starts to drop. This is the perfect moment for the switch to be closed again.

So what this regulator does is to send the current in small packets, first into the capacitor and then from there a much smoother current is sent into the battery (if you are interested in the details, the best thing will be to buy a book about 'switching regulators'; this fascinating topic would be by far too technical for this book). The regulator measures the voltages over the capacitor and the battery plus the average current coming from the panel. From this information the perfect ratio between the time during which the switch is closed and the time in which the switch open is derived. The data of the coil determine the switching frequency. That's how we generate the perfect voltage for charging the battery.

The switch in the circuit is by no means a mechanical one (as in the old cars) but a MOS-FET. This type of transistor has the big advantage that even a very small amount of energy is sufficient to swap states - from conducting to blocking and vice versa. We remember $P = V * I$. When a transistor is conducting there is only a very small voltage drop over it; therefore nearly no energy is converted into heat in the transistor. When a transistor is blocking then the current is zero and again no heat is produced. Heat is only generated during the short time when the state of the transistor is changing. Most PWM regulators work with frequencies higher than 20 kHz in order not to molest the human ears (the changing current will cause the coil to deform slightly so that it swings, which is why transformers hum). The relatively small energy loss in the transistor is acceptable, the coil itself can be small (coils are expensive) because of the relatively high frequency and the ripple on the voltage delivered to the battery is minimal. A good solution!

Well designed PWM regulators have an efficiency of 90% and some as much as 97% or even 98%. The efficiency here is defined as the ratio between the energy piped to the battery on the one hand and the energy converted into heat within the regulator on the other.

12.5 Energetic differences: linear and PWM regulators

The first phase of charging a battery is called bulk phase charging. We try, without damaging the battery, to get as much energy into it as possible. This is achieved by charging with the highest permitted current (with a 100 Ah battery this would be 10 Amps; for more information read the supplementary chapter about batteries). The output voltage of the panels must be high enough to force such a current into the battery.

We assume that the panels are big enough, so let's take two examples. In the first one the panels have an output voltage of 24 Volts and in the second one it will be 16 Volts. We neglect the voltage drop over the regulators. We assume that the current into the battery would be 8 Amps at a voltage of 13 Volts.

A linear controller with 24 Volts input voltage

> Between the input voltage and the output voltage is a difference of 11 Volts. So we have a power loss of 8 A * 11 V = 88 Watt.

A linear controller with 16 Volts input voltage

> Between the input voltage and the output voltage is a difference of 3 Volts. So we have a power loss of 8 A * 3 V = 24 Watt.

A PWM controller with 24 Volts input voltage

> Into the battery goes 8 A * 13 V = 104 Watt. The regulator has an efficiency of 90% so there will be a power loss of 10 Watt.

A PWM controller with 16 Volts input voltage

> Into the battery goes 8 A * 13 V = 104 Watt. The regulator has an efficiency of 90% so there will be a power loss of 10 Watt.

The PWM regulator seems to be absolutely superior. Additionally, in order to get rid of 88 Watts of power loss with the linear regulator, one needs quite a big heat sink. However, now comes a really big surprise: on an energetic level there is no difference in efficiency between the two types of regulator. Surely this can't be, haven't I just proved the opposite? Then carefully follow my ideas.

The linear regulator works continuously and dissipates the energy not needed for the battery in the form of heat. So far everything is clear. However, what does the PWM-regulator do? During a few milliseconds it takes the energy from the panel and then it switches this energy flow off. So what does the solar panel do during the time that the switch is open? Nothing other than . . . . getting warmer!

The only energetic difference between a linear regulator and a PWM regulator is the place where the heat is produced - in the linear regulator the heat is produced inside the power transistor and the PWM regulator produces the heat within the panel. As far as efficiency is concerned there is no difference at all! The claim of producers of PWM regulators that this type of charger had any advantage over the linear regulators is simply flannel. The only real difference is that the PWM regulator uses the big solar panel itself for heat dissipation while the linear regulator uses a large aluminum body with cooling fins.

Nowadays simple PWM controllers which can deal with quite a few Amps are sold for something like 10 Euros. Since finned aluminum heat sinks cost more, the PWM controller is to be favored because it is cheaper (but not better).

12.6 The MPPT regulator

When discussing the function of solar panels we had already taken a look at the Maximum Power Point (the MPP). I made up a table displaying how much power could be drawn from a panel if resistors with different values were connected to it, and then left that topic. Now it is time to unearth it again.

I start with a drastic example why it makes sense to look for something better than a PWM regulator. Let's assume that we have bought a panel which can deliver 180 Watts at 36 Volts; at its MPP it will deliver a current of 5 Amps. We also assume that we are using a PWM regulator in order to charge our battery.

Since within a given circuit the current must be the same no matter where we measure it, it means that no more than 5 Amps can be pushed into the battery. For the sake of simplicity we assume that the battery needs to be charged at 13 Volts again. The result is that only 65 Watts (5 Amp x 13 volt = 65 Watt) goes into the battery. Officially a typical PWM regulator has an efficiency of 90% so there will also be 6.5 Watts heat dissipation loss in the regulator itself. That's not a major problem.

However, we also know (see above) that with this type of regulator the heat is dissipated elsewhere, in this case from the solar panels. On a nice day with clear skies (visibility from pole to pole, as pilots used to say) the panel could theoretically put out the 180 Watts printed on the label at the back.

The regulator and the battery take 71.5 Watts (65 Watt pushed into the battery plus 6.5 Watt efficiency loss); that means that the remaining 108.5 Watts are converted into heat dissipated by the solar panel. The battery would be quite happy to take all of the 180 Watts that the panel could put out but in practice 115 Watts (108.5 + 6.5 = 115) are turned into heat one way or another. So we get an approximate loss of 64% which in turn means that we have only 36% efficiency. Since the panels are still the biggest cost in a solar installation, it can be easily seen that money is thrown out of the roof hatch (near to the panels) by using unsuitable charge controllers.

A professional installer should have learned how not to design an installation, but I bet that there are lots and lots of smaller solar installations out there designed by semi-professionals and enthusiasts who did not know how to make a proper design. In fact, how could they know which parts of their installation don't match properly to each other? They don't have ways of finding out that the panel is two degrees warmer than it should be. After all, which amateur knows how to calculate the current going into the battery from the technical data? Who is able to do so? You! Now!

Well, this was a pretty long introduction and I would not have written it if there were no remedy. Its name is . . . . the MPP regulator! I first explain the technical idea behind them.

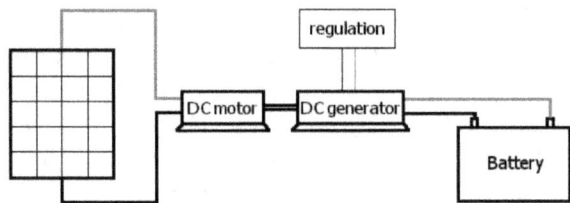

In this picture we have a solar panel which is directly connected to a DC motor. This motor is coupled mechanically to a DC generator which feeds into our battery. The output voltage is regulated by the voltage supplied to the exciter coil (hopefully you remember that we had something identical when talking about old cars). The regulator measures the battery voltage and then calculates the perfect current for charging.

In these old-fashioned cars the power going into the batteries was relatively small compared to the power of the main engine (in fact, the driver regulated the amount of current going into the battery by use of his right foot on the throttle, without being aware of it). Now the ideal situation would be that the power output of the generator is identical to the power output of the panel at its MPP. The regulator needs to take into account the technical data of the panel (we come back to that in a moment).

If the regulator thinks that the power flow into the battery should be higher then it increases the voltage on the exciter coil; the output voltage of the generator increases and the power pushed into the battery increases as well. Since the power has to come from somewhere the generator throttles the motor which in turn changes its inner resistance. In this way we can, with a given irradiation, move along the characteristic curve of the panel just by changing the voltage at the exciter coil of the generator.

When the regulator requests a huge amount of power then the generator slows down the motor drastically and the resistance of the motor goes up towards infinity. That means we don't get any energy into the battery. If the regulator demands nothing then the motor is freewheeling and we don't get any energy into the battery. Now we face the $64,000 question: how can we manage the regulation in order to get an optimum amount of energy out of the panel?

A few pages back I drew up a table which showed how with different resistors we obtained different power outputs from a panel. In this diagram here (source: Wikipedia, Stündle, public domain) we see two current curves in red for two different irradiations (the diagram is 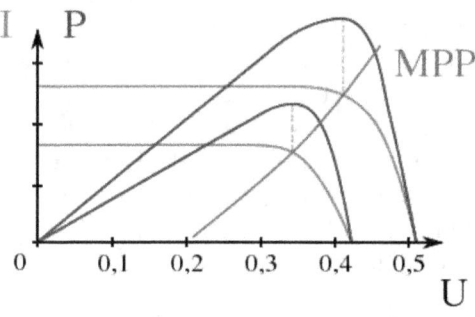 for a single cell), and in blue we show the electric power so generated. Using different resistances in the consuming device makes it possible to force the output voltage down. Then we can go up to the corresponding blue curves and read the power delivered from the panel.

Which red line is used is determined by the intensity of the actual irradiation (so there is a blue line for every red line). When we connect all the points on the red lines at which the corresponding blue line is at its maximum we get the gray line; this is the so-called MPP line. So now we come to the justified and pragmatic question: "And so what?" Actually, I have to admit that in this form it is of nearly no use unless we know the actual irradiation and all the technical data of the panels, after which we would have to get all this information into the regulator. But indirectly we do get some benefit: the gray curve is increasing in a strictly linear fashion! It is with this bit of information that we can do something.

We make the regulator just a little bit intelligent so that it can work by using trial and error. We just start with a relatively small amount of power (if we want to charge a 500 Ah battery using 5 Watts there is no sense in starting any work at all). If we can't reach that level at present, we just wait and try again a few seconds later. Shortly after sunrise we will get our 5 Watts. Now the regulator increases the voltage at the exciter coil slightly so that we could possibly get 10 Watts instead. If we can't get them, we fall back to 5 Watts.

There is a wide field for developing algorithms which can find the optimum as fast as possible but for us it is sufficient to know that we take the following steps (the "=>" means "leads to"): output of a new exciter voltage => a change in the power going into the battery => a change in the load on the motor => the motor displaying a different resistance to the panel => a move along the curve of the actual irradiation to the working point. When a change of the exciter voltage causes a decrease of the power flow into the battery we simply change the direction of our search. So we are swinging, with small steps, around the perfect working point. The blue curves in the diagram (giving the power output) have a pretty flat summit; that means that even if we are slightly beside the optimum the amount of power not gained can be neglected.

The absolute fantastic thing about this method is that when the irradiation changes this regulator changes automatically to the next current curve (by going up and down the gray line) and directly goes to the MPP point on it. This enables us to get the most energy out of the panel at any moment, and this also explains now the 'T' in the title of the chapter. It stands for Tracking. We have a Maximum Power Point Tracking. So this regulator follows the MPP to wherever it moves (if you knew all the panel data and you could measure the actual irradiation you could do without the 'T'; but tracking is much simpler).

This design of such a regulator is absolutely brilliant. The regulators can be manufactured without any knowledge of what the actual installation will be. The desired result, to get as much power out of the panels, is achieved automatically, and it is the MPP curve (the gray one) which makes that possible because it is increasing in a steady and linear fashion.

Normally nobody will build such a regulator using DC motors and generators (though it would be possible). Instead electronic components are used to achieve the same result. In order to make things as easy to you as possible just think of a MPPT regulator as a special DC to DC converter where the input resistance can be changed just as the output voltage can be changed (these two are really independent).

The result of this complicated circuit design is that we have two distinct circuits. So we have an input circuit defined by $P_{In} = V_{In} * I_{In}$ and we have an output circuit defined by $P_{Out} = V_{Out} * I_{Out}$. Because of the law of the conservation of power P_{In} and P_{Out} must be identical (we neglect losses).

The disadvantage of the PWM regulator (and the linear regulator) was that the current in the input circuit had to be the same as the current in the output circuit. Therefore the power transmitted varied depending on the required voltage. With a MPPT regulator you have a continual power loss of roughly 10% but this loss is much smaller than when using a PWM regulator (unless, purely by chance, it happens to work at the MPP). We need to have (again!) a closer look.

12.7 Comparison: PWM regulator vs. MPP regulator

We already saw that the PWM regulator is quite a clever gadget because the power losses are produced within the solar panel and not in the regulator itself (that's where the super-duper efficiency values come from). Therefore I'm treating it like a normal linear regulator. The efficiency of a linear regulator is calculated with the formula:

$$\text{Efficiency}_{\text{Linear Reg.}} = (\text{Output Voltage} / \text{Input Voltage}) * 100$$

If we have an input voltage of 24 Volts but we need 12 Volts at the output, then we get an efficiency of 50% which is abysmally bad. If we have an input voltage of 18 Volts and we still need 12 Volts at the output, then the efficiency increases to a fantastic 66% (which is still pretty lousy). One can simply say that the efficiency of PWM regulators (or linear regulators) being used as charge controllers in photovoltaic installations becomes worse with higher voltage differences between input and output. The fact that the MPP is only used by pure chance makes the efficiency of such a regulator even worse (yes, that's possible; but what is the comparative of abysmally bad?).

Ultimately a MPPT-charge-controller is nothing but a DC-DC converter where the ratio between the input and the output voltage can be adjusted by help of a controller. That is, the output voltage can be both lower as well as higher than the input voltage. Basically the input to the regulator is the absorbing capacity of the consumers (which includes the battery to be charged as well). Then the resistance seen by the panels is adopted in order to receive exactly that amount of energy. Often it will be necessary to look for the actual MPP but if the panel could produce more power than needed, a working point other than the MPP is selected. If more power is needed than can be delivered from the panels then the output voltage is reduced.

In our actual example the MPP controller is 20% more effective than the PWM controller. Now we have to make an economic evaluation because PWM controllers start at about 10 Euros whereas MPPT controllers are (still) nearer to 100 Euros. We have a look at a small system which delivers just 1 kWh per day costing 7 cents when working with a PWM controller. We assume that the installation will work well for 20 years, and during this time will deliver a total of 7,300 kWh, worth 511 Euros.

If we had replaced the controller by one that was 20% more efficient then we would have gained additional energy worth 100 Euros. In terms of cost per kWh there would be no difference. This situation changes the moment when the price is not 7 cents per kWh but 15 cents, so it is not really possible to decide which kind of controller makes more sense. It depends on the voltages and how much any kWh will cost (sorry, but you will have to do your calculations on your own).

However, if you have a look into a MPPT controller you will be quite surprised. There are only a few electronic components more in the box than in a PWM regulator (for sure the program in the little micro controller will be much more sophisticated). In the long run the MPPT type of controller will win because it has the potential to become a lot cheaper (I would bet the farm that we will see MPPT controller in the 200 Watt range for 20 Euros before 2020)!

12.8 Addendum: direct connection / shunt PWM

In chapter 12.1 I had written that this type of regulators is pretty unsuitable because it is difficult to find a good match between panel and battery. Well, there is one exception. You can directly connect panels with 36 cells (open circuit voltage 18 Volts) and 12 Volt batteries or panels with 72 cells with 24 Volt batteries. The output voltage of the panel will be pulled down by the battery and the working point of the panel will be slightly below the MPP.

The higher the SOC (state of charge) the nearer the working point moves to the MPP. The only (and absolutely important) trick with this kind of regulator is that the connection between panel and battery must be interrupted when the battery voltage comes up to the point where it starts to gas (there are some variations which start with some kind of PWM to keep the battery voltage on the preservation level). With a real minimum of hardware these regulators deliver a pretty high efficiency.

You should only play with the idea of using such a regulator if you do know exactly what you are doing. If there is a mismatch between panels, regulator and battery than you will most probably have to replace the regulator by a more expensive MPPT regulator.

One word more: only use this kind of regulators in small installations!

13. Modifying solar installations

When planning a completely new installation one has the advantage that one can select (more or less) freely which components to use. In other words you have more money than you really need because most people would be quite happy if they could pay just for their actual energy need. The usual case will be that they start with a small installation.

For our next example we just look at the owner of the nice little house in the mountains of Andalusia; in order to make things a bit easier we will call him Jan de Witt from the South of Holland. When he bought the piece of land with the house he was told that there would be no problem at all to get a connection to mains electricity. He could even earn money with it if he sold permission to connect other people to his power line.

At a distance of a few hundred meters he can see the mains cable; in one way quite near but in another way as far as the other side of the galaxy. After quite a lot of effort he found out that indeed it was possible to bolt a transformer to one of the masts down in the valley in order to supply him with electricity, but when he was told the price he felt a bit dizzy for a moment. Until he realized that you can't get a civilized life without paying the price.

Then he found out that none of the owners of the land between him and the target mast was willing to allow the cable to cross their land or even willing to talk about permitting Jan de Witt to set masts on their land. After quite a bit of talking and thinking he found a way that he could put the cable and the masts on the boundaries of the plots and the owners agreed to that plan. Now the cable was 850 meters long instead of 600 meters but some sacrifice must be made if you want a civilized life. Then he asked for the price for setting the masts and mounting the cables and for a short moment he had the impression that the world darkened in front of his eyes.

After this he applied for official permission, since in a country like Spain you can't just set up power lines as you like. After a while he got the impression that his application had dropped behind the desk whenever it was time to make a decision about giving approval. While chatting with one of the neighbors he found out by chance that her second cousin was working in the municipal administration and would have a look at why there was no reply.

The result of this investigation was that Jan's house was indeed built with official building permission but that the local council never ever had the right to give this permission. Since the house was finished (there was a roof on top) it was 'quasi' legal but he would never ever get a connection to the grid. For a staggering three-figure price a lawyer told him that there was no way to prove that the seller did know about this. The world in front of Jan's eyes seemed really dismal.

In the evening, down in the valley at the bar (where Paco, the owner, allows him to charge his laptop and mobile phone), Jan got to thinking about his problems and only one thing became really clear: life without electricity is simply not on! At the very least he would need a steady 12 Volts supply in order to charge his phone and the laptop, but where to get it? The Internet seems to provide an answer to every question; next day he would look up what was said there about alternative energy.

He spent an absolutely frustrating day in the Internet; within five hours he found out that there is a solution and that the solution is not even very expensive, but more than that he could not find out. The different forums were really interesting but to about one thousand questions he had he only found three to five answers, and in exchange he had the feeling that these answers generated 20 to 30 new questions each. This was no way to get ahead, it was just taking too much time! Maybe a good book could be his salvation?

With most books sold via Internet it is possible to read at least a few chapters in order to find out whether the book is roughly what you are looking for. After about one hour of looking at books Jan was prepared to continue with the forums. But not tonight, perhaps tomorrow (if you look up the word 'mañana' in a dictionary you will usually find it translated as 'tomorrow'; but this translation is pretty wrong since the real meaning is 'no way, not today'. I personally think that 'on any one of these wonderful days' is the correct translation). Then, just by chance, he spots the book "Photovoltaics for the Technically Ungifted, Academics with Two Left Hands and all Others". The sample chapters are quite readable but on the chapter about physics he really had to concentrate (his time at secondary school is long past).

If it was really true that the content would not become more complicated than the chapter on physics then it would be worth a try; and at the price of 8.95 Euros you can't go wrong a lot. Just half an hour ago he wanted to walk down to Paco's bar down in the valley in order to get drunk as a lord, but now he downloads that book and decides to go there by car. Tomorrow he will have to concentrate. With this in mind he grabbed his flashlight and blew out the three candles on the table.

I introduced this story about Jan de Witt so that you can understand how most privately owned installations get started unless an architect makes the full planing beforehand (NB the basics of the story are true but happened near to Valencia). First it is essential to get some lighting in the house and you don't think about anything going any further. Then it would be nice to be able to charge the laptop and the phone so that it is not necessary to ask Paco all the time. Well, Paco said that you have a lifelong right to charge your computer in his bar but it is perhaps better not to have to depend on this promise. Then you buy a car radio with CD-player and eventually comes that really big plasma TV!!?.

So, after a while it is time to buy another panel or two. In order to avoid technical problems you start to look for the type of panel you bought two years ago. Which is of course exactly the type which is not produced any more. After some research you find a similar panel (of the same size and which would fit mechanically) but which would deliver 120 Watts instead of 80 Watts. Is it possible to connect the new panel to the existing panels without electrical problems? And if so, how? I'm going to answer these questions in this chapter (and tell a bit more).

13.1 Aging of panels

In 1977 a few universities in Germany (and most probably in other countries as well) started doing some research on the long term effects on a variety of solar panels; in the course of the research quite a few different models and types were added. One result of this research was that panels do get older and do lose power. The next important result was that deterioration with age became less and less due to technical advancements (currently the estimated lifetime of panels is far more than 20 years unless they were produced on a Monday morning). The third main result was that the complete failure of a panel is very rare (or rather, if we're being very technical, the power that any given cell can deliver may get smaller and smaller but it hardly ever happens that the cell fails completely; that would in most cases mean that the panel as a whole fails).

The actual state of the art (in 2013) is that panels still retain 90% of their power after 10 years and after 20 years they still retain 80%. Some manufacturers even label their panels with less peak power than the panel actually can deliver under STC, in order to claim a higher power output after ten years, but for the end user this does not matter all that much.

If I plan a solar installation which exactly meets my personal needs today then I must take into account that in a few years time I will lose a small percentage even if my needs did not increase. So the question of how to enlarge my installation will show up almost automatically in the course of the next few years. Even without plasma TV.

13.2 Connect voltage sources in parallel

The symbol generally used for a voltage source is shown in the little diagram. Often the '-' and '+' symbols are missing. So I use a simple memory aid: the line of the plus pole is longer, so it could be cut into two halves to make a plus sign while the shorter line is just long enough for the minus. Other DC sources can be displayed correctly with the same symbol.

The most important value of a voltage source is the nominal voltage. That is why this

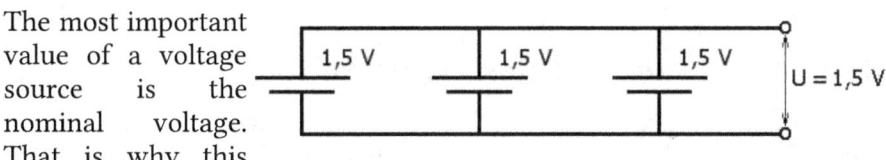

value is normally set directly beside the battery symbol. In the picture we have three voltage sources and all of them have the same nominal voltage. If you connect them in this way in **parallel** then you will have, with the same nominal voltage, a higher capacity (that means that you might leave the light switched on for three hours instead of the one hour with just a single battery).

In the same way we can connect solar panels to each other if they have exactly the same nominal output voltage. If we connect the panels with a nominal output voltage of 18 Volts then the charge controller might need only two hours in order to charge the battery instead of the six hours it needed when connected to just the one panel. If you take the panels one by one and measures the output voltage then there could be small differences. If the differences are less than 10% (with identical panels the difference will be much smaller) then there is no problem. This is because the charge controller forces down the output voltage anyhow in order to come to the MPP (or at least near to it).

It will be something completely different if you interconnect two big 1.5 Volts batteries in parallel with a small 4.5 Volts battery. First the 4.5 Volts battery will charge the two 1.5 Volts batteries and then the three batteries will discharge at the same speed. While the 4.5 Volt battery will already be in the red (danger) section the two 1.5 Volt batteries are still in good health. When the two 1.5 Volt batteries come into the region of danger the 4.5 Volt battery will have been dead for a long time.

Solar panels can take more (you may remember that for them all states between a short and open contacts is normal and they don't get damaged no matter how long that situation lasts). However, if I connect 9 panels with 36 Volts output voltage in parallel with a panel which delivers only 18 Volts I would not like to make a bet (unless that one panel is fitted with a protection diode).

Principle: interconnections in parallel only with identical output voltages!

13.3 Voltage source interconnected in line

In the next picture we have four voltage sources interconnected in line (NB the unit at the bottom of the drawing provides 4.5 Volts!). As you can see the various individual voltages add up to 9 Volts.

However, now we come to the problem caused by this connection. Let's assume that all the batteries each have a capacity of 100 Ah except for the battery with 4.5 Volts which only has a capacity of 1 Ah. As long as the circuit is open there is no problem but what will happen if we connect a consuming device which uses 9

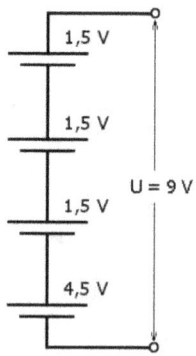

Volts? Well, after a very short time the 4.5 Volt battery will not be able to deliver more electrons and the other batteries still have 99% of their capacity. The three batteries will bundle their power and force a current through the 4.5 Volt battery (that means that they try to reverse charge it; if you are lucky you only have to replace the 4.5 Volt battery but probably you will need to mop up the problem).

As I already mentioned, solar panels can take a lot (hopefully you remember the chapter about light and shade). If there is no free wheeling diode then the maximal current is determined by the weakest panel; more can't be delivered. So if you have an old panel in line with new panels the output power will be less than what you might have expected but no damage will take place. You can check for weak panels by measuring the short current. If the current of that one panel is much smaller then don't use it any more (or put it in parallel with other panels). The panel with the lowest short current limits the power output of the complete string!

Principle: interconnections in line only with panels of the same capacity (in Ampere!)!

When interconnecting batteries or solar panels in parallel we will get a higher capacity (capability to deliver work, kWh, stored energy) at the same voltage. When we interconnect batteries in line we get a higher capacity (capability to deliver work, kWh, stored energy) at a higher voltage.

13.4 Reducing line losses

We just go (mentally) back to the chapter where I cheated a bit regarding the cable cross-sectional area. There we had an installation which was delivering 1 kW at a voltage of 25 Volts. From $P = V * I$ we got the information that we needed 40 Amps in order to pass that energy along the cable.

When reading that part of the book you could not know that it is also possible to connect panels in line (unless you learned that somewhere else). I just copied the tables with the resistances, the permitted currents and the costs here so that you don't have to scroll back.

mm²	Ω per meter	Amp. accord. DIN	€ per meter
1.0	0.0178	11	0.12
1.5	0.0118	15	0.17
2.5	0.0071	20	0.26
4.0	0.0044	25	0.42
6.0	0.0029	33	0.63
10.0	0.0018	45	1.05
16.0	0.0011	61	1.70

First, we just connect two panels in line (that is called a string) and all strings in parallel. By doing so we have 50 Volts on the cables leading down from the roof. With 50 Volts we only need 20 Ampere in order to pass 1 kW. That means we can take the cable with 2.5 mm² which has a resistance of 0.142 Ω if 20 meters long. A current of 20 Ampere will cause a voltage drop of 2.84 Volts over the cable and that means we have power losses of 56.8 Watts in the cable.

Now we connect four panels in line and the voltage increases to 100 Volts (DC voltages higher than 120 Volts are considered dangerous; so we are still on the safe side). At 100 Volts we only need 10 Amps for the transmission of 1 kW. For safety reasons we select the 1.5 mm² cable which has a resistance of 0.236 Ω. A current of 10 Amps produces a voltage drop of 2.36 Volts over the cable. That means we reduced our losses on the cable drastically down to 23.6 Watts.

Before the thinnest cable we were permitted to use had a cross section of 10 mm² and we had losses of 57.6 Watts. Now we have a cross section of just 1.5 mm² and the losses are down to 23.6 Watts. To put it simply: losses on the line decrease with the square of the voltage. Hence it is just a question of a few trials (and some errors) to find out which voltage and which cable you should select in order to have the highest benefit. The prices for cables you know (at least roughly) so you have now all the information you need.

(Lets assume you decided to replace your old PWM controller by a new MPPT controller. To rewire the panels on the roof is an easy job to do; but why not reuse the thick 10 mm² cable? It has a resistance of 0.022 Ω. A current of 10 Amps produces a voltage drop of 0.22 Volts over the cable. That means we reduce our losses on the cable dramatically down to 2.2 Watts. You will have to admit that this really is an improvement).

In the example a few pages back we had losses of 57.6 Watt with the thinnest 'legal' cable. That caused financial losses during ten years of €210.24 and with the new solution we reduced them down to 8 Euros. If the MPPT regulator does not cost 200 Euros more than the PWM regulator it might be financially interesting to kick out the old regulator.

In commercial installations voltages of up to 1,000 Volts are common (not only because of the line losses; the inverters for direct feeding to the grid get simpler as well), but this technological solution is forbidden to us donkeys. All voltages higher than 120 Volts DC (and 50 Volts AC) are potentially dangerous (and may even be deadly), and even when supplying electricity for a big household a tension of 120 Volts will be far more than is really needed.

13.5 Charge controllers and higher voltages

In the data sheets of charge controllers you can look up the maximum input voltage. I don't know of any linear or PWM regulator for 12 Volt batteries which could cope with a 100 Volts input, which means that what I have to say in this chapter is more or less an academic exercise (such a charge controller would be technically no problem but the higher the voltage a transistor can handle the higher will be its price; that's one of the reasons why no such regulators are produced, but we are now going to have a look at the real reason).

Our solar panels could be delivering 10 Ampere at 100 Volts, which is 1 kW. At the battery end we receive 10 Ampere at 12 Volts, which is 120 Watts. That means we would have an efficiency of 12% and losses of 88%! No matter whether it would be a linear regulator or a PWM regulator, a device with such poor efficiency nobody would want to buy.

However, an MPPT regulator will have an efficiency of more than 90%, no matter how high the voltage is. So either you have to do without higher voltages (then you could use a PWM-regulator and thick cables) or you buy a MPPT regulator which can be quite a little bit more expensive.

An essential advantage when working with higher voltages is that the interconnection of the panels becomes much easier. If you put four panels in line only two cables come to the junction box instead of eight. In order to interconnect four panels using MC4-connectors only four

MC4-clips have to be pushed into each other (so that the panels are already connected) and you need one extension cable to connect the last panel in the row to the junction box. Now you have to connect two cables in the junction box and the job is done. If you prepared the additional cables beforehand you might need 30 seconds per connection and there is much less cable on the roof (and it is much easier to keep a beady eye on the system).

13.6 Light and shade again

What was said a couple of pages before about solar cells is true for complete panels as well (but it all depends a bit on the internal connections). We stick to the example with 4 panels delivering 25 Volts each.

We connect four of them in parallel and we are about to cast a bit of shade on one of them. When doing so we measure the voltage and the current using a suitable load (for example lorry head lights). Now we cast some shade on one of the panels and we see that the power output varies between 75% and 100%.

Now we connect the panels in line and we use a suitable load (maybe the heating element of a washing machine). If we now vary the shade over the one panel the output power varies between 0% and 100% (I am just repeating something which you should already know).

The best solution would be to install your panels in a place where there is no shade at all, but often you will not be able to arrange things in this way. In such a situation you should plan your panel connections in such a way that as few strings are affected at the same time as possible. In some cases it might be even necessary to use more than one charge controller in order to allow different areas to be fully independent.

Just to make it all clear: if you use just a single MPPT controller in a situation where the panels have different shading conditions, it will search for the MPP of all the panels taken as a whole. If you have one area with shade on the panels and a different area without shade then the working point will not be perfect for any of the areas; in such a case it would be advantageous to use two regulators.

There are situations where it becomes impossible for a simple algorithm to find the maximum power point. If you use MPPT controllers which can under all circumstances find the MPP then you will have to pay the price for that. So try to keep the situation simple so that you can use simpler (and cheaper) devices; each part of the array with a different irradiation condition gets its own controller. The professional installer in your neighborhood might come up with much better - and much more expensive - ideas.

13.7 Several charge controllers to one battery?

There could be two reasons why you may want to use more than one controller. Firstly, your installation has grown and the solar controller has become too small (a perfectly normal occurrence) and secondly you may have to deal with a complex situation with shading and you have to segregate different parts of your installation (as I mentioned previously, you then are perhaps reading the wrong book).

From the view point of the controller it looks as if it gets electric energy from the solar panels and on the other side is a battery to be charged. If you, while the controller charges the battery, switch on any consuming device, then to the controller it looks as if the battery has a higher capacity since it obviously takes longer to charge it. If there is another controller feeding into the battery then the controller decides that the capacity of the battery has become smaller because it can be charged faster.

With good chargers the charging of a battery consists of at least three phases. First there is a phase with constant current (we want to fill the battery as fast as possible). Then, when the charging voltage comes near to the voltage at which the battery starts to gas strongly (details in the supplementary chapter on batteries), we have a phase with constant voltage (with the current becoming lower and lower). Finally, when the current falls below a threshold the charger switches to a voltage which allows it to keep the battery in a fully loaded state (it just evens out the self discharge).

Just in order to keep things straight: if you happen to own a charge controller that is unable to follow these phases it is best if you drop the controller into the bin (or give it as a present to your dearest enemy). The same applies if your controller is not capable of falling back into a previous phase. There are controllers out there that seem to think that there is nothing in this universe other than themselves and a single battery. If you own one of them and you wish to keep it, then please use this book to start your barbeque. Right now.

Normally, charge controllers are designed in such a way that they can co-exist, without any problems, with other charge controllers. However, some charge controllers seem only to be able to live in peace if the other charge controllers come from the same manufacturer. Which of these two situations will be true is probably not something you will find in the data sheets. In principle one can say that the good charge controllers (well, they have the tendency to be a bit more expensive) will make no problems whatsoever. The collaboration of several charge controllers is a standard task!

So if you own panels with different orientations (East-West roofs have become very popular of late because you get a power input which is much more constant over the course of the day) or the electrical data of the panels are not compatible then the easiest way is to use different charge controllers. The few Euros you spend for a second controller is next to nothing compared with the hours spent in order to find another good and reliable solution.

In addition, any aid in finding complicated solutions would contradict the motto of this book: photovoltaics are simple! So best we leave it with the statement: if you have to enlarge your installation, deal with it as if it were a new (and separate) installation (no, you don't need more batteries, I'm just talking about panels and charge controllers).

From this advice there is one exception: the supporting framework! This framework should be as big as might eventually prove to be necessary (and by eventually I mean in 20 years time or so). Then it becomes very easy to install a few more panels whenever it seems necessary. A bigger framework only costs very little more than a framework that is made to measure. Whereas if you have to extend the supporting framework it will be complicated and much more expensive. So if you are a single person just now you should plan the supporting framework in such a way that a family of four could live with it.

Before I close this chapter I would like to pick up two topics: the equalization of batteries and the mixing of the battery liquid by letting the battery gas (the details you can find in the supplement chapter about batteries). In both cases a voltage is connected to the battery (or battery bank) which is normally too high. The battery reacts on this voltage by splitting the water into hydrogen and oxygen (not all the water at once but milliliter by milliliter).

Now we simplify the situation by assuming that we have two charge controllers being connected to the same battery in parallel. Both charge controller deliver a voltage of about 13.2 Volts and the battery is near to full. One of the chargers decides (or you switch on the function) that a voltage of 14.5 Volts should be given to the battery so that it starts to gas (whether this happens in order to mix the battery liquid or for an equalization does not matter at all).

The voltage over the battery is not what the other charger expected and the only reasonable thing to do is to behave as if it were not there. That could be realized in the way that the power output is disconnected (by help of a relay) until the battery voltage is again in a range which makes sense to the charge controller. Or (a better solution) it follows the output voltage and adjusts the output current to zero, until it thinks that a higher voltage would make sense (so it will never pull the voltage down).

In Germany we call this principle: whoever shouts loudest is right and the other follow!

If one of your chargers does not work like this then buy a new charge controller!

14. Storing electric energy

Until now we have only looked at the question of how to generate electricity using photovoltaics and how much this costs. As far as a diesel engine is concerned, it does not matter at all whether it is running during the day or at night, but a solar panel can only generate electricity while the sun is shining; during the night there is nothing. So we need to think about how we can have electricity during the night as well as during the day and what to do if a depression darkens the sky for a couple of days.

For our immediate needs it is enough to worry about the question "How much does it cost to store one kWh and use it again at a later time". So what we need to know in a storage system is the number of possible storage cycles (the number of times we can store and retrieve the energy), the efficiency (the energy losses caused by storage) and the investment costs per kWh; from this data we can calculate the cost per storage cycle. What we are not going to look at is the amount of space we need for that storage since most of these solutions are out of our price range in any case.

Type	Max. No. of Cycles	Efficiency	€ / kWh	Cent / Cycle	Cent* / Cycle
Super Capacitor	500,000	90	10,000	2.22	101.36
Supercond. Coil	1,000,000	90	2,000	0.22	20.09
Fly Wheel - Iron	1,000,000	90	5,000	0.55	50.22
Fly Wheel - CFK	1,000,000	95	1,200	0.12	10.95
Lead-Acid Batt.	1,000	80	100	12.50	-
Pumped Storage	10,950	80	71	0.81	-

(Numbers taken from German Wikipedia; I think that the cycle of a lead-acid-battery is cheaper).

With the pumped water storage power plant I assumed that it lasts 30 years and runs one storage cycle per day. If we assume one storage cycle per day, then with the first four methods listed we would come to about 2,740 years of usage. In the column „Cent* / Cycle" I based my calculations on a lifetime of 30 years only.

Even if we leave out all the possible difficulties (the space required, accessibility, permissions needed), the first three solutions are of no interest to us. Only the CFK flywheel is better than the lead-acid battery from a cost point of view. This is also the reason why flywheels are used in electric distribution networks for stabilizing purposes. Small disturbances are very frequent on the grid, so in this application the full number of storage cycles can be used.

Lead-acid batteries and flywheels can deliver electricity within milliseconds and free all the stored energy within a short period of time. However, flywheels have one big disadvantage: they need a steady feed of about 1% of their stored energy in order to overcome the losses through friction.

Naturally there are many kinds of electric accumulators (batteries) and the question then arises of how to make an economic decision. One criterion could be their environmental impact. What you think about the different technologies and how they should be rated I leave to you; but I would like to take up a cudgel for the lead acid battery which has not really deserved its bad reputation. This type of battery has, within the EU, a recycling rate of over 95% and the percentage of the recovered lead is very near to 100% (only gold has a better recycling rate).

Probably the most important criterion will be the price of a full storage cycle of 1 kWh of electricity. If we divide the price of a given type of storage by its capacity then we can standardize the purchase cost (with batteries we obtain the capacity in kWh by multiplying the capacity given in Ah with the nominal voltage; a battery with 100 Ah and 12 Volts has a capacity of 1.2 kWh; and this number is then reduced to the advised maximum discharge for that battery type).

Additionally we need to know how long the storage system will last when used in a sensible way. For that we need to take the data sheet of the battery and look at how many deep cycles it can take. Then we multiply the capacity by the number of deep cycles and get how many kWh we can run through the battery within its lifetime. If we divide this number by the price of the battery then we get the cost of a single storage cycle of one kWh (we leave out for the moment its efficiency and maintenance costs).

Now we know the cost in Euro per kWh and the question of how long our storage will last (in years or days) has been eliminated. So it does not matter whether the battery lasts 3 years or 7 years, we know the costs per kWh stored and extracted. Let's have a look at three different battery storage types.

1. Small NiCd Accumulator
    - Price: 1.35 Euro
    - Voltage: 1.2 Volt
    - Capacity: 1.3 Ah
    - max. cycles: 300

    - Capacity: 1.0 * 1.2 V * 1.3 Ah = 1.56 Wh = 0.00156 kWh
    - Throughput: 300 * 0.00156 kWh = 0.47 kWh
    - Costs: 1.35 Euro / 0.47 kWh = 2.87 € / kWh

2. Camper van / Solar Battery
    - Price: 199.90 Euro
    - Voltage: 12.0 Volt
    - Capacity: 180 Ah
    - max. cycles: 600

    - Capacity: 0.8 * 12 V * 180 Ah = 1.73 kWh
    - Throughput: 600 * 1.73 kWh = 1,038 kWh
    - Costs: 199.90 € / 1,038 kWh = 0.19 € / kWh

3. Traction Battery (6 PzS 840)
    - Price: 3,390.00 Euro
    - Voltage: 48 Volt
    - Capacity: 840 Ah
    - max. Cycles: 1,500

    - Capacity: 0.8 * 48 V * 840 Ah = 32.562 kWh
    - Throughput: 1,500 * 32.562 kWh = 48,843 kWh
    - Costs: 3,390 € / 48,843 kWh = 0.069 € / kWh

As you can see it is pretty easy to determine the utility value of different types of battery in a solar installation. Differences vary between 6.9 cents and 2.87 Euros for a full storage cycle of 1 kWh. At this stage I have left out the efficiency in order to make the calculation easy. I will now take it into account for the two lead acid batteries.

I will just accept, for the moment, the advertisements of the sellers/manufacturers which state that the maintenance-free batteries have an efficiency of 90%. If that were so, we would come to a price of 21.1 cents per full cycle of 1 kWh with the solar battery. The PzS battery is often labeled with an efficiency of 80%, so here the price would come to 8.6 cents. In fact, the maintenance-free battery is 2.45 times as expensive, per storage cycle, as the battery that needs a little maintenance. Perhaps I was a bit unfair to compare the big 'primitive' battery with a medium size 'advanced' battery but even so the result could be that you might decide to do a little bit of maintenance (just topping up with water). As reward you get electricity during the night for less than half the price!

With closed (but not sealed) traction batteries we have to remember that they must be stored somewhere (because of the gassing action this type of battery must not stand in habitable rooms unless you use special stoppers which are interconnected by help of a tube which leads outdoors). We also need other material (fuses, switches, cables, meters etc.) so that eventually we come to the conclusion that you will have to pay something like **10 Cent per full storage cycle of 1 kWh** (some nerds tell that they came down to about 5 cents). If you have already calculated how much each kWh generated during the day will cost then you can work out the price per kWh used at night. Depending on local conditions (i.e. the price per kWh of electricity from the mains) you can make up your mind whether it is better to run your own independent system all the time (you will then need to buy yourself some bigger batteries) or whether it pays to switch to mains at night.

When we have a look at electricity prices we see that the kWh is traded on the so-called spot markets for 6 to 7 cents in Europe (though this price is calculated on megawatt hours). At times, because there was more than sufficient alternative energy, spot prices were negative (!!!), i.e. the industrial consumer was paid for using up electricity. However, normal price differences are far too small to make large scale battery storage an economic option. The lead-acid battery has been a well known technology for many decades and so it would be surprising if this type of battery became cheaper (although if demand rises I would not be surprised to find that prices were to increase slightly).

Li-ion accumulators are still at the long term development stage; in 2013 it was estimated that this technology will need another 10 years to break the 10-cent-per-kWh barrier. From this technology no solution is in sight (at least not in the near future).

To carry out electrolysis and later generate electricity using fuel cells has an efficiency of round about 30%; large scale storage is possible but is not cheap. When converting electricity (using chemical processes) into natural gas and later back to electricity, the efficiency will be roughly as low as using fuel cells. But (at least in Germany) there are huge storage capacities and the gas pipeline grid is fully developed (there is very little problem in transferring almost any amount of energy from A to B).

On the electricity market a concept has arisen which is called 'contribution to profit'. The principle is that anybody generating electricity would like to sell each kWh for at least 6 cents. If the price goes down to 4 cents the point is reached that it is, from an economic point of view, the same whether to sell electricity per kWh for this price or to convert the electricity to gas in order to generate electricity at a later stage and to sell it for a price above 6 cents (gas power plants have a short reaction time so that they can quickly cover peak demands; one might get 9 cents if there is no need for haste and the gas is not going bad).

On the other hand, there are other people who think that PGP installations (Power to Gas to Power) will never pay for themselves since they are not running all day long and the biggest part of the costs is not running the converter but paying back the original investment.

Be that as it may, Germany will be forced to undertake another pioneering task. Maybe unwillingly, but the consequences would be far too expensive if that job is not done properly. That will all become very interesting!

15. Inverters

Up to now we have covered subjects such as how to generate DC from sunlight and how to use a charge controller in order to store the power in a battery (and there is no reason why we should not use some of that energy while the batteries are being charged). In essence we have everything we might need - in the appliance market for camper vans and yachts you can find practically everything you might possibly want to have and you can find 12 Volts types or 24 Volts versions to suit your system).

That's fine in principle but the solutions leave a lot to be desired when you have a look at the prices and compare them with the prices of appliances designed for 230 Volts and 50 Hertz (please insert here the typical values for mains appliances in your country). The 12/24 Volt versions are far more expensive! In this chapter we have a look at the possibility of converting DC to AC so that we can use cheap mains voltage appliances.

The simplest way (and it was actually used some decades ago) is to couple a DC motor directly to an AC generator. Such a set will show quite a few kilograms on the scales and you would have to renew the bearings every few years but such an installation is perfectly robust. As prices of electronic components decreased and electronic engineers learned more during the last four decades, nobody is making such devices any more.

However, such a device delivers a big didactic advantage because it makes it simple to explain what happens in principle. You know what a DC motor does and you have at least a faint idea what an AC generator can do. What the engineers and technicians did was to exchange such a set of electrical machines to a box full of electronics.

And the engineers and technicians did their job properly and they designed nice little gadgets which you can buy, ready to operate, in shops or via the Internet. On one side DC goes in and on the other side AC comes out. The inverter controls all the voltages, currents and frequencies internally (that is, if it is meant for isolated operation, and we will come back to that shortly) and if anything is not exactly as it should be then the inverter swaps to the passive mode and behaves as if it were not there (this reminds me of Harry Potter and real magic). Or to put it absolutely simple: after connecting the inverter there will be absolutely no reason to check or to do anything for years and years - compared to cheap batteries inverters are really maintenance free (absolutely boring!).

15.1 Feeding inverters

This type of inverter feeds electricity directly to the electricity company's mains grid. In order to do so it first has to synchronize with the grid and thereafter the inverter tries to support the grid as well as it can (which means that a few hundred Watts try to support quite a few Terawatts!). This support is mainly concerned with the form of the sine wave and the frequency of the AC voltage (what exactly happens inside the inverter you don't really need to know).

Any inverter feeding into the mains must conform to quite a lot of regulations. I just list the three most important ones:

> 1. The design must make it impossible for disturbances to be passed on to the grid.
>
> 2. When the grid shuts down the inverter must immediately stop working.
>
> 3. The inverter must be able to compensate for a certain amount of reactive power.

Regulation 1. This refers mainly to the form of the output of the inverter. The voltage should be a nice sine wave. Anybody who has taken a careful look at the wave form of the voltage in the grid (especially inside or near an industrial area) will have grave doubts whether that wave form can still be called a sine. The meaning of this rule is that an inverter must at least not worsen the situation. Inverters feeding into mains must deliver electricity in the form of a sine wave with little distortion (more about that soon). This is often referenced as a pure sine wave.

Regulation 2. This demand is connected with the operational safety. The output of an inverter is 230 Volts AC and any contact with AC higher than 50 Volts can be deadly. Let's assume that the electricity company has to do some work on the cable near to your home, so the electrician will disconnect all the cables. Normally he should now make sure that there is no tension on any of the cables and then shorten and earth each one. However, what if he forgot? What if he did not know about your installation (because you forgot to tell the electricity company that you installed one)? Then he is in mortal danger. That is why all inverters which feed into mains need to check every few seconds whether there is still what is called a trailing voltage in the grid. If not, they have to shut down their output until there is a strong sine wave on the line again!

Regulation 3. This is a bit difficult to explain. On the line we have a tension in the form of a sine wave. Some electric devices take in some energy at the start of any half cycle and push it out at the end and there are other devices which do it vice versa. Thus there is some energy swinging back and forth between generator and consuming device a hundred times per second. This amount of energy is called blind energy because while it doesn't actually perform any work it does provoke losses on the line; the electricity company will have to pay for these losses and they don't like it one bit. With new installations the inverter must be able to compensate for some amount of blind energy (so that the blind energy is only swinging back and forth along the short cable in or near the house between the inverter and the appliance and not over the long cable between the inverter and the generator of the electricity company). So the owner of the inverter has to spend some money so that the electricity company can save some more money. Not nice, but what can you do?

Lately in Germany some additional regulations were added. Large installations need to have some way of regulating the power output. This regulation is either done directly by the electricity companies (they can instruct the solar installation to reduce the power output by for example 50%; if they do so they have to compensate the income so lost by the owner of the installation). Such a regulation can be done automatically in case the grid gets out of control. A second thing is that the inverter must measure the voltage and the frequency. Depending on these values the inverter has to reduce its output or disconnect completely (disconnect if the voltage level is above or below specified levels; reduction even down to zero when the frequency becomes too high).

This intervention into the private electricity production was technically necessary because every now and then the amount of electric energy from photovoltaic systems or wind turbines changes strongly without any previous warning. As a rule of thumb you can say that the bigger the rotating generator set, the slower the reaction to changes. Nuclear power plants can't intervene when rapid changes occur even if they wanted to do so.

Ideally, the owner of the grid would like to be able not only to reduce the output of photovoltaic installations but also to regulate the consumption of the appliances in households as well. The idea is called the smart grid and you can read more about it in the supplementary chapter on that subject.

15.2 Isolated inverters

When we are talking about an island installation (or an isolated installation) we mean an installation which has no contact at all with the electricity grid; it could literally be on an island. Whereas a peninsula is an installation which can either work autonomously or be connected to the mains. A typical peninsula will work autonomously whenever the weather is nice and during the day. At night or when there are long times of bad weather the installation will draw electricity from the mains. A variation of a peninsula installation is one where most of the energy from the panels is used directly and only the surplus is fed into the grid.

The wave form of the voltage is a sine and cycles will normally be exactly 50 Hertz (this could be different in your country). In the European compound grid there are some thousands of power plants working in a synchronized fashion. So if you connect an inverter meant for isolated installations to the grid it might work with 51 Hertz instead of 50; if you connect such an inverter to the grid you can only hope that the fuses react much faster than the electronics (although I would not like to bet on that), so probably you will get some smoke signals. If your inverter is feeding the net and supplying power for internal consumption at the same time you will need an inverter which is fully synchronized with the grid.

An uninterrupted power supply, the UPS used in many a small office, is for all practical purposes some kind of peninsula. You normally get the electricity from the grid but if that fails then your own electricity production starts automatically (the electronics of inverters are so fast that they can take over within a few milliseconds), the office net is disconnected from mains at all poles (for reason of safety) and the inverter starts powering the office net. Often this switching will be done using relays so that there is a safe galvanic separation (neither DC nor AC can leave the house in direction of the grid or vice versa).

Such a switching will take about 10 to 20 milliseconds (mechanical inertia) and with most light sources you might notice a short flicker. Single phase machines (vacuum cleaner, hand mixer etc.) will equalize the extremely short blackout through their inertia. All electronic appliances (including PCs) have got internal power supplies with a capacitor big enough to bridge this short time (many UPS are used for computer installations!). Inverters of the 'luxury class' are at all times synchronized with the grid in order to avoid possible spikes of voltage or currents when switching.

15.3 Inverter technology

The electronic inverter is far from being a new invention; it has been continually developed over the last few decades. Depending on the power needs and the situation where they are used there are different ways to build them.

In the next picture we got the wave form AC voltage should normally take: a pure sine. The sine is defined by help of the unit circle (circle with radius 1) which means therefore the maximum value is 1 and the minimum value is -1. The voltage in the European compound grid is 230 Volts; strictly speaking, this is not the maximum value but the effective value in that a DC voltage with this value would produce the same power. You get the peak value if you multiply the effective value with 1.41 (that is the square root of 2), i.e. 230 Volt * 1,41 = 325 Volt. The 1 at the y-axis corresponds to this value of 325 Volts.

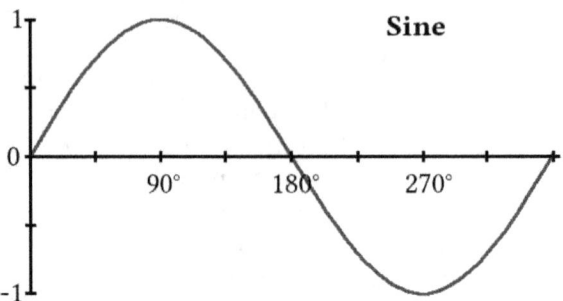

A rotating generator will automatically create such a wave form, as the coil rotates within a static magnetic field. Using only electronic circuitry it is difficult to generate a real pure sine and the simplest approximation is the square wave (one can proof mathematically that any cyclic signal can be produced by help of the first harmonic and all the other harmonics; the sine is the first harmonic and in a square wave we have loads of higher harmonics). With some appliances this does not matter, but other devices (for example hifi music systems) do not function really well with such a voltage square wave form (you might hear the loud screech of a saw or a moped revving up).

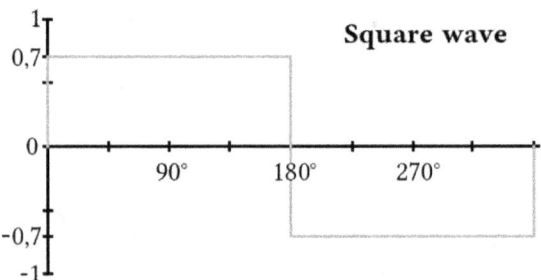

The easiest way to create a symmetric square wave signal is to have a positive and a negative voltage and two transistors which are conducting and blocking as a simple push-pull system. Whenever a sine would cross the zero line then the transistor which was conducting will start to block and vice versa. In order to prevent a short (i.e. both transistors are conducting for a brief instant) there is a very short period of time when both transistors are blocking.

Then somebody had the bright idea to make this very short period a bit longer. In this way the fraction of higher harmonics is reduced considerably. I assume that some sales person then came up with the idea to call this wave form 'modified sine' (well, at least that sounds much better then 'square wave with reduced disturbance voltages').

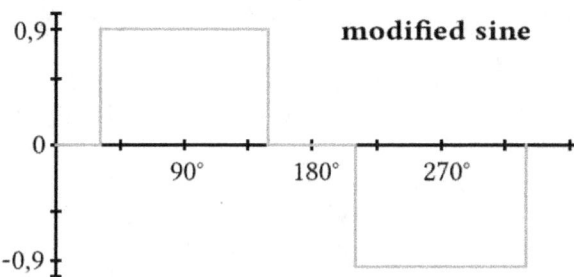

All three wave forms have the same effective voltage (the area between the x-axis and the signal has the same value with all three signals). This is achieved by adapting the maximum voltage.

With this modified sine wave many more appliances can work without major difficulties than with the square wave. However, other appliances such as many laser printers, scanners and modern coffee machines will not work at all or make difficulties (Murphy's law will make sure that this is going to happen when you definitely don't need problems). If the inverter has to be as cheap as somehow possible this wave form is the best you can expect.

15.4 Working principle of a pure sine inverter

As nearly always in technics and electronics there are different ways of reaching the goal and they will differ (normally considerably) in price and robustness.

First of all there are those inverters that use a transformer and others that do not. The advantage of an inverter with a transformer is that one can have a galvanic separation between DC and AC side (electrons can not flow from one side of the transformer to the other side, i.e. even if something goes wrong, you can't have 230 Volts AC on your panels).

Normally in installations one wants a 'one fault safety'; transformers are seen as faultless devices (they do fail but that is that seldom that it can be ignored). If the voltage on the DC side is kept below 120 Volts than all work done there is save even with the installation running.

Additionally there are high frequency and low frequency inverters (both with or without transformer). The high frequency has the advantage that the transformer can be pretty small and it might be possible to deliver an almost perfect sine. The advantage of the low frequency inverters is that they are much more robust (you will find them in high power installations); because of the bigger transformer they weigh a lot more and often cost more.

I'm now going to explain how a low frequency inverter with a transformer (for separating the AC and the DC side) works in principle. In order to keep the losses low we will use a toroidal transformer with one primary winding and some secondary windings (if that is not telling you anything please use the search facilities in the Internet or just skip the complete rest of this chapter).

Normally only AC voltages are applied to a transformer but it is perfectly alright to use a DC voltage as well. When you connect a DC voltage to the primary winding then it is acting as a coil and a coil tries to resist the current by building up a magnetic field or by reducing it (the strength of the magnetic field increases/decreases in a linear fashion). Because the magnetic field changes in a linear fashion the voltage induced into the secondary windings stay constant so we have different DC voltages there. How high these DC voltages are depends on the ratio of the windings between the primary winding and the secondary windings. If you have on the primary side 30 windings and on the secondary side 60 windings then the output voltage will be twice as high as the input voltage.

If we leave the input voltage connected to the primary winding for too long a time then the iron core will go into saturation (the strength of the magnetic field does not change any more) and the output voltages will drop to zero. In order to prevent this from happening we use a circuitry as shown. Here we see four transistors

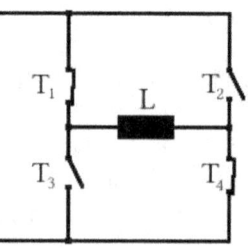

and L is the inductor of the primary winding of the transformer. The transistors T_2 and T_3 are conducting when T_1 and T_4 are blocking and vice versa. The nice thing with this circuit is that the direction of the current is reversed every time the transistors change state. If we do that a hundred times per second then we have a 50 Hertz square wave signal on the secondary side which is symmetrically to the zero Volts line.

How we can make good use of this is something we can see in the next picture. Here we have two voltages V_1 and V_2 with V_2 being twice as high as V_1. These voltages we could get from a transformer as just described. The output voltage is now generated by either connecting V_1 or V_2 or both in line with the output. The result is still a bit rectangular but it already looks much more like a sine then the modified sine.

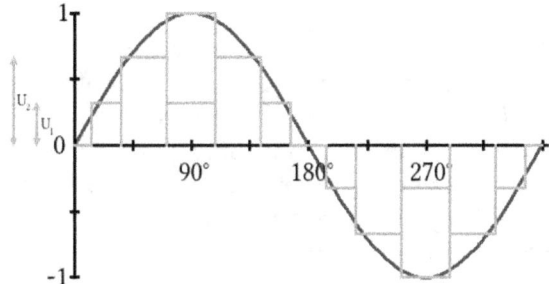

If you are a fan of good music you will most probably have heard the term 'distortion factor'. This term describes the difference between an original signal and a copy - it is the percentage of the difference in the areas of the two signals and the x-axis. Simple music equipment will have 10% distortion factor at full output whereas in Hifi equipment it must not exceed 1%. You need a very well trained ear to be able to detect 0.5% distortion.

What does that mean as regards our inverter? Well, clever people computed that if you have 8 different voltage levels the distortion factor will be between 1% and 4% (depending on when you switch); in order to realize these 8 steps you only need three voltages and a way of adding them. If you use 5 voltages and a system of adding then the distortion factor will be down to between 0.2% and 0.4%, so you couldn't hear it any more although one could measure it (with good equipment).

If our toroidal transformer has 5 secondary windings delivering the following voltages (10, 20, 40, 80 and 160 Volt) then we could expect a wave form at the output which is near to perfect.

Such an inverter would be relatively heavy (with a continuous power delivery of 3 kW it would weigh between 25 and 40 kg). A transformer has a life span of far more than 20 years and it would cost about 50 Euros if build in a country with low wages or using highly automated production lines. Additionally one needs a circuit that computes which voltage should add to the output and which not; such a board might cost 20 Euros. Additionally we need the boards which deliver the voltages and add them up (each costing maybe 15 Euros); even somebody with no knowledge about electronics would be able to repair such an inverter just by swapping printed circuits (if he has a small testing device it might even be very fast).

I guess that inside Europe there is not a single manufacturer willing to bring such an inverter onto the market (where would be the point if there is no high profit?). Such a Jeep under the inverters could be sold in the developing world for decades (in some parts of Asia and South America lots of Jeeps produced during the second world war or shortly afterwards are still being driven around; in India they are still being produce though with a different motor).

If the design of the inverter is such that it can cope with overloads (lets say during 400 milliseconds) of ten times the nominal load and will still work when the temperature is 50°C and the sun shines on the case, then the manufacturer will not be able to cope with the demand (unless he makes the price that high as if it were high tech from the best).

If you are interested in further information you might look for the dissertation of Sérgio Daher (June 2006, Kassel, Germany; ISBN-10: 3-89958-236-5 or ISBN-13: 978-3-89958-236-9). You can find it as well in pdf format in Internet.

At the start of the book there was a reference to an article from India about the prices of inverters. Based on the dissertation of Sérgio Daher I calculated the prices and I think that one could undercut the prices in India (from 2013) drastically (shot from the hip I think that 70 Euro per kW are possible). That would be the technology the world needs! A market nearly without limits! Just as an inspiration for the industry (in China?).

15.5 Which inverter to buy

When it is time to buy an inverter you will have to choose one out of a thousand or so different makes. In the end you might end up using a counting-out rhyme, throwing a dice or tossing a coin. But before we come to that, here is a list to work through (maybe afterwards you have to trust your luck but not right now).

a) Power
Most important is the maximum power output of the inverter. Sometimes it is given in Watt or Kilowatts and sometimes you will find VA or kVA; think of VA as Watts (that is not strictly correct but a good enough approximation for making a decision). This number limits how many and what sort of appliances you can switch on at the same time. The minimum you need is defined by that appliance which consumes most, after which you have to decide what more you want to use at the same time. In a normal household a 3 kW inverter will do quite well.

b) Power Peaks
We were talking about the very high start-up currents when motors are switched on (for example in fridges and freezers) and that there is actually nothing you could do about this fact. Such high currents typically last about 200 Milliseconds. If you buy an inverter which can cope with ten times the nominal power but only for 20 Milliseconds that is no real help (normal inverters can cope with twice the nominal power for a few seconds and that is often no help either).

c) Slow Start
Some inverters lower the output voltage when the current becomes too high. At times this could be practical, for example when you connect a washing machine to an inverter. Then the strength of the current does not jump but goes up gradually. Disadvantage: there is no guarantee that your washing machine (or whatever you want to use) will work with a slow start inverter. And don't try to use your PC at the same time when you run the washing machine; it might want to re-start every time the

spin dryer starts (this is the fault of the inverter and not of the washing machine!).

d) Stand-by / automatic start-up
If your TV is in stand-by it is not switched off; parts of the electronics are still working (years ago the cathode-ray tube - remember cathode ray tubes? - of the TV was preheated so that you only had to wait 15 seconds instead of 3 minutes until you got a picture on the screen). The inverter does not know at what time during the night somebody switches on a light or the fridge motor starts, so it is idling (and using 20 to 30 Watts, basically doing nothing). The other method is to generate a pulse every few seconds in order to detect whether an appliance needs power and then to start working (this method uses 3-5 Watts while waiting). Have a look at the size of your battery and the price of the inverters, and then make a decision.

e) Noise
There are many reasons why an inverter generates noise. If it uses a large transformer it might hum loudly, while other inverters might cheep. Even if the noise level is really low (i.e. less than 35 dB) such a noise can drive you mad when you are trying to sleep. The same applies to fans. Some fans are running all the time emitting unpleasant sounds. Others are switched on when the internal temperature rises and then there are those that regulate their speed according to the temperature - these normally just whisper. If there is a 30 cm concrete floor between the inverter and your sleeping room you might not care. If you have such an inverter in a camper van you may possibly decide (from the second night on) to disconnected it before you go to bed.

f) Remote control
If the inverter is in the house in which you live, a remote control will probably not be something you really need. If the inverter is built into a camper van (and it has to go in a well ventilated place which you find hard to reach) you might be happy to have one (uncivilized people simply disconnect them from the battery).

Please remember: you have all the time the option to make separate circuits!

So you might have one 'modified sine' inverter for the kitchen (including the washing machine). They are relatively cheap and normally all kinds of motors should work without any problems (but remember, I give no guarantee!). Since you want to feed the fridge and the freezer with it, it should have a low consumption when in stand-by mode, since it has to run 24 hours a day (and it must be able to cope with the fact that the compressors of fridge and freezer start to work at the same time!).

You could have a circuit with a 'pure sine' inverter for the fully electronic coffee machine, the Hifi equipment, the TV and your computers (or other delicate and expensive electronics). This inverter you could switch off in the evening.

Additionally you might have a circuit for the lighting. Whether you do it on a 12 Volt level (use a DC-DC converter if you have a battery bank with more than 12 Volts; else you will ruin at least one battery within a short time) or on 230 Volt AC level does not matter too much if inverter/converter have a low stand-by consumption (you might have to live with the fact that you use the switch and it takes 20 seconds before the light is on; switching off will be with normal speed).

If you worked your way through this list your decision might not be perfect but it will definitively not be bad (if it comes to tossing a coin: select tails!).

16. Overview: electric motors and generators

There are two major types of electrical machines: motors and generators. Motors take electric energy and transform it into mechanical energy. Generators take mechanical energy and transform it into electric energy. In practice (nearly) all generators can be used as motors as well and (nearly) every motor as a generator. That is why we can group them together, by and large, as electrical machines.

Next we distinguish the type of electric current used in the machine. So we have electrical machines using DC and others using AC. In the power range of a few Watt there are practically only DC machines. In the range of a few hundred Watt there are machines of both types. The power range starting at about one kW becomes the (nearly) exclusive domain of AC machines.

The regulation of small DC machines is somewhat easier than the regulation of AC machines. But over the last 30 years the knowledge of how to do it properly has increased a great deal and nowadays this difference is negligible.

We can find large numbers of small electrical machines in cars and commercial vehicles (starting with the electric windows and finishing with the ventilators). In this segment DC machines are mass produced and they are therefore relatively cheap (but some manufacturers seem to forget that these machines - especially when they are sold as spare parts - are made of iron and copper and not of copper and gold).

The standard electrical machine consist of a stator (that is the part which does not rotate; normally the stator is integral with the housing) and a rotor (the rotating part of the machine). The rotor generates a magnetic field which, relative to the rotor, is static. The stator generates a magnetic field which, relative to the stator, is rotating.

In a generator the magnetic field of the rotor tries to push the magnetic field of the stator (the rotor is propelled mechanically from the outside) and in a motor the rotating magnetic field of the stator tries to push the magnetic field of the rotor. Maybe you remember from school that like magnetic poles repel each other and unlike magnetic poles attract each other; this, together with rotation, is the basis of all electrical machines. So the difference between motor and generator is simply the question of what pushes and what gets pushed. This makes it possible to use the same electrical machine in a pumped storage hydro power station to pump the water up to the storage or to generate electric energy when it flows down again.

Within the AC machine group there are two major types which differ in the way the magnetic field of the rotor is generated. They are called asynchronous machines and synchronous machines. Asynchronous machines are mostly found in the lower power range and synchronous machines are used at the top end of the range.

Synchronous machines act as if the static magnetic field of the rotor and the rotating magnetic field of the stator are coupled together, rather like the gears in a car gear box. Up to the limit that would break the gears in a gear box the magnetic fields move synchronously (which is where the name of this type of machine comes from). If you have a synchronous generator and you want to feed the electricity it produces into the grid, you first need to get the rotor to revolve at the correct number of revolutions and then the machine is connected electrically to the grid (rather in the way old cars start moving: first you push the clutch pedal down, then you ignore the screaming of the cogs and finally you release the clutch; car speed and motor rotations are now coupled).

Asynchronous machines are more like cars with an automatic gear box. You step on the gas (that is you switch on the electric machine) and with the engine running at full power the car starts to accelerate. This is managed by allowing slipping action between the speed of the two magnetic fields (how exactly this works is not very important at this stage).

If you own a car with a moment-to-moment indicator of fuel consumption you will have observed the following effect. Driving at a steady speed, the car might use 7 liters per 100 kilometers but when you change down and push the gas pedal to the floor the consumption jumps up to 30 to 40 liters per 100 km. That is exactly what happens when an asynchronous motor is switched on. Most asynchronous motors need, for the 200 millisecond start-up sprint, ten times the nominal power they need when running steadily, and there are even monsters out there which will not start to work if they don't get 20 times the normal quantity.

With machines working with AC there is one more distinction, and that is the number of phases; there are machines using one phase and there are others using three phases. When AC was first developed for electrical machines, the same type of current was also intended for lighting streets and houses. However the most important aspect was the need to transmit power for generating motion. Here, at the power station, was a big steam engine and there, a few kilometers away, was a production facility (for example with drilling and milling machines), which needed to be powered.

The problem was to transfer a rotation from 'here' to a rotation 'there'. So they found out that the easiest and cheapest way was to use three phase AC. Three phase AC? Let's start by having a quick look at the basics.

If you have a metal coil and you let a magnet rotate in front of this coil, the changing magnetic field in the coil will produce a voltage, and this voltage changes in the form of a sine wave. A full circle has got 360° and a full cycle of the sinus is defined over these 360° as well (unit circle). Now if we take three coils and position them around the rotating magnet (one coil at every 120°) then in each of the coils a sine wave is induced but shifted by 120° relative to each other. We now take a big step and connect the three coils of the generator directly to three corresponding coils of an electrical motor. Now comes the big surprise: no matter how we turn the rotor of the generator back and forth, the rotor of the motor will follow this movement without any hesitation or delay.

Instead of transferring all the power over one big cable it is transferred over three (smaller) cables and without any further effort we have the close coupling of generator and motor. As always in the area of engineering: the cheapest solution that solves all problems wins.

This solution is so good that it is used all over the world for transferring electric energy. Synchronous generators, up to a few hundred Megawatt, are installed in the big power plants. In a single grid all these generators are directly coupled as if they are all part of a single gear box. One could say they all rotate together as if they were one single machine. So, if you are going to change something in the electric installation in your house which is not protected by a fuse, then making a short will not end in the same way that the fight of David against Goliath finished. Most probably some of your cables will vaporize into a metallic cloud and the grid will not be impressed in the slightest way (this possible consequence is another reason why work on such installations must be done by fully qualified technicians).

Now we come to the big question "What has all this to do with my photovoltaic installation?" Well, first of all, you should be interested in the environment in which your installation works. The second (and more important) reason is that single-phased asynchronous motors are very cheap to produce and therefore are used a great deal; on the other hand they cause a lot of problems in photovoltaic installations.

The (nasty) behavior of asynchronous motors is well known to design engineers. The engineer's job is to make sure that current going through the machine is not too high, which destroys the machine, and to make sure that the fuses in the fuse box don't blow. On the other hand the machine has to be as cheap as possible because that is where the profit of the manufacturer is made. Engineers are well aware of these problems, and so they make sure that the machines don't cause problems when connected to mains. But that is exactly the environment that we haven't got in a photovoltaic installation - we simply don't have gigantic power reserves.

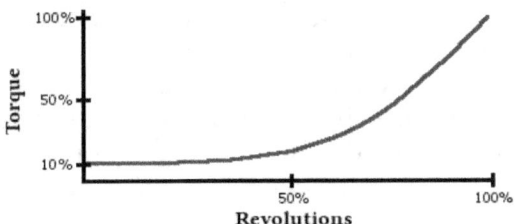

In this picture you see the typical diagram for this kind of motor as fitted to a huge variety of fridges and air conditioners. When the motor is switched on and has not yet started to turn, it only produces about 10% of its potential torque. That in itself would be no big problem but attached to the motor is a compressor - and the engineers designed the motor in such a way that it can just start with this additional load. To start an electrical motor coupled directly to a compressor will need ten times as much power (for the first 200 Milliseconds) than it uses when running steadily (the Watts usually marked on a little plate at the back of the appliance).

There is simply nothing you can do to change the behavior of these machines because the current they draw is necessary for them to start rotation from any possible starting condition. If you insert a slow starter circuit in order to limit the current at the initial moment it may well be that the current is not sufficient to start the motor in some situations. If that happens the motor does not start rotating but starts to heat up until the thermal security switch (at least there ought to be one!) disconnects the power supply. Without such a thermal switch the motor might actually start to burn (even thermal switches might fail).

Then these motors have another bad habit. They consist mainly of copper cables wounded around an iron core; these are the coils. One could say that coils are extremely tenacious. If you connect a coil with a voltage they resist the current by pumping the energy into a magnetic field. When you then disconnect the voltage the coil uses the energy in the magnetic field in order to continue to push the current. An open contact has a very high resistance but the coil does not care; it raises the voltage until a current flows (the coil has no other means of getting rid of the energy in the magnetic field). You then need a new inverter (or somebody who knows how to repair them and they are a rare species nowadays).

One possibility for protecting the electronic circuit of the inverter from these voltage peaks (most inverters are not designed to power bigger motors; they normally have a protection circuit and usually this might be sufficient - the problem is that sometimes they are not) is to install an over-voltage arrester. That sounds like a big deal but is in fact just a small component which costs about one Euro and is called TVS protection diode.

This little element is assembled directly between the two 230 Volt terminals of the motor. A suitable TSV protection diode is the 1.5KE400CA; the CA stands for bidirectional use (this is important for us since the motor is driven with AC); the 400 stand for 400 Volts (at higher voltages the diode behaves like a throttled short). After the diode has absorbed the energy from the magnetic field it behaves again as if it were not there (that sounds familiar, does it?).

An aside: When working with 230 Volts AC, this number gives the effective voltage (that is, a DC voltage of this value would deliver the same power). The voltage at the peaks of the sinus is 1.4 times higher, that is 325 Volts. Since the voltage in the grid can (legally) be up to 10% higher, then 400 Volts is the minimum threshold voltage the diode needs to have. If the threshold voltage is much higher then the protection is not all that useful any more (and if it is lower the diode will work for a very short time like a fuse and will offer no protection in the future)

16.1 Why is this a problem with photovoltaic installations?

Lets look at a standard electric drill, to be found in most households (I'm talking about a simple one without electronics). It has a nominal consumption of 500 Watts - that's what it says on the label. When the motor rotates freely (i.e. when we are not actually drilling a hole) it will consume less than 50 Watts. When you start drilling the consumption goes up and when you press harder you can hear the motor making a tortured sound and starting to slow down then really a lot of energy is used. Depending on the make of the drill, it might draw up to 10 kW when the motor starts (not the modern ones with electronics). At 1.000 Watts the small inverter you are (probably) using is waving the white flag of surrender; for safety reasons it switches off (it would be an even worse situation if the inverter checks every two seconds to find out whether the - from its point of view - short still exist; I wouldn't like to bet on how long the drill and/or the inverter would survive).

In fact, asynchronous motors are exactly what we don't really want connected to our photovoltaic system, but unfortunately they are built into all machines designed for domestic use. The reason is simple: when connected to the 230 volt mains taken from the grid they don't cause problems and the specialist market where their behavior causes problems is too small. So in the future you will have to ask again and again for products that are compatible with solar installations! Maybe one fine day a manager might think that he has found a new and interesting market. Then you will get what you need.

Machines and equipment for yachts and camper vans does not form a big enough market. There are products for 12 Volts or 24 Volts but these are really expensive. This might be because the owners of such vehicles are known for having a lot of money when it comes to their hobby. The know-how to produce electric motors for photovoltaic installations is there, but what is missing is a market which justifies these products becoming mass produced (a typical chicken-and-egg problem).

16.2 The working principle of heat pumps

In order to understand how a fridge or an air conditioner works, you have to understand the principle of the heat pump (the picture was made by Ilmari Karonen and is Public Domain).

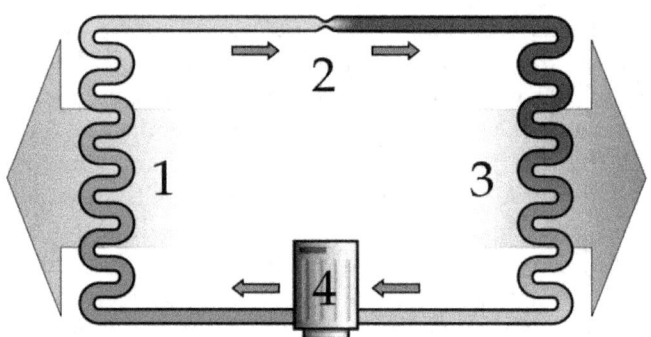

We use electricity to run a compressor (4) in order to heat a volume of gas by increasing the pressure. It cools down in the condenser (1) and becomes a liquid. All the red area is under high pressure. By help of a restrictor (2) this pressure is reduced, which produces the desired side effect that the liquid becomes very cold. In the evaporator (3) the cooling agent absorbs heat from the surrounding. (1) and (3) are in fact heat exchangers.

In this circuit a cooling agent is used which can be either liquid or gasiform within the limits of the pressures and temperatures. The cooling agents used in earlier years were not really environmentally neutral. Nowadays propane is often used in fridges. This gas can burn and in a correct mixture with air it is even explosive, but the actual amount of (liquid) gas in the fridge would easily fit into a cigarette lighter. These fridges are not excessively dangerous.

Now we come to the point why this working principle so interesting. Heat as well as electric energy can be measured in Watts and with a heat pump one can get 4, 5, 6 or even more thermal Watts out of one Watt of electric energy (this is called the Coefficient Of Performance - or CoP - in very technical publications; the heat pump of a fridge will have a CoP of about 3). The question is: what is the working principle?

Lets assume we have two lakes near to each other; the surface level of one lake is 20 meters higher than the other. If we want to pump water from the lower lake up to the higher lake then we will need to use a certain amount of energy in order to do so; if we allowed the water to run back to the lower lake, we could use a turbine to recuperate most of this energy (although, owing to friction we will lose some energy). Now let's suppose that the lower lake is well protected from the wind and the water heats up considerably when the sun shines on its surface; we don't really want that to happen since we are running a fish farm in the lower lake and when the water is too warm the fish don't grow very fast. That's why we decide to pump the water up to the upper lake, where a fresh breeze cools the water.

So the evaporator (3) corresponds to the lower lake where heat is absorbed and the compressor (4) is in fact a pump. The condenser (1) is the upper lake and the restricter (2) corresponds to the turbine. In the picture of the two lakes on thing is missing until now and that is that the water heats up when it is compressed, so the water reaches the upper lake with a temperature being higher than the water of the lower lake. And when the water runs down to the lower lake again, the uplift pressure is decreased and the water running into the lower lake is cooler than the water in the upper lake. And exactly this effect produces the Coefficient Of Performance (CoP).

The working principle of a heat pump also explains why an air conditioner can not only be used for cooling but also for heating. The connections of the two heat exchangers (1 and 3) are switched over - for a heat exchanger it does not matter whether it absorbs heat or dissipates heat. That's all that is necessary to turn an air conditioner into a heating unit.

16.3 Fridges and solar energy

In many parts of the world fridges are a standard appliance. During the last few years there has been a considerable change in the technology of fridges (not a change in the main principle but more in the details). Fridges changed from being power hungry carnivores needing a lot of electricity to cuddly little kittens that sipped delicately. In 1994 energy efficiency classes were introduced within the EU. For lots of different appliances, in many sizes and types, standard virtual devices were defined that consumed standard amounts of energy; appliances were then graded in various classes from 'A' to 'G'. 'A' means that the appliance uses much less energy than the virtual device and 'G' means it uses much more. The technical changes we mentioned earlier have made it possible that within the EU no new fridges are produced or sold with a classification worse than 'A+' and it has even become necessary to expand the scale to 'A++' and 'A+++'.

This classification must be stated on the label attached to the fridge in addition to all the other necessary information. What we are interested in is the actual consumption and the standard states that this must be given in kWh per year. Suppose I pay €150 for a simple fridge with classification 'A+'. This fridge has a capacity of 92 liters (no freezer compartment) and a consumption, per year, of 117 kWh. This means that the fridge only consumes, on average, 0.32 kWh per day. If we assume a warmer environment we may need to increase this value to 0.5 kWh per day. With a price of 22 Cent per kWh (as in Spain) we will have to pay 11 cents per day to run the fridge.

When one looks for gas driven fridges you may be surprised: they are pretty expensive (I presume that is due to the fact that they are mostly used by the owners of camper vans who are willing to spend a lot of money on their hobby). Within the EU you can assume that a gas driven 60 liter fridge will cost about €500 (if it is driven by 12 Volts then the price will be even higher). Such a fridge consumes about 300 grams of butane per day; butane was costing €1.20 per kilogram in Spain in 2012. That is, we have daily costs of 36 cents for the gas. The fridge costs about four times as much to buy as a normal 230 volt fridge and the daily costs are three times as much.

Now the question is: how much will it cost if we buy a normal 230 volt fridge and power it using solar energy? I will carry out this calculation for a stand-alone solution so that we can observe the full impact of the various costs. We know that the average consumption is 0.5 kWh per day but we do not yet know the consumption when the fridge is actually running (well, when the motor+compressor is running). If you wish to know more about how to find this out, please read the supplementary chapter on measuring consumption.

Modern fridges consume, when the compressor is actually running, between 75 and 100 Watts (I'm not talking about big commercial fridges for restaurants or hotels). If you are still using the fridge you inherited 20 years ago from your aunt Gertrude (it brings back so many happy memories) then the consumption might be nearer to 400 Watts and the compressor will be running for much longer periods. Use a consumption meter and then decide whether these memories are really worth such an amount of money (year in year out) or whether there are cheaper ways of thinking about her.

So we don't use the old fridge from aunt Gertrude but we buy a new modern one. In order to provide it with power we will assume that we can do so directly from the solar panels for the 10 hours a day that the sun provides power, and for the other 14 hours we need to store the necessary energy in a battery (in the supplementary chapter on batteries you will find the details of why it costs about 10 cents to store - and use subsequently - 1 kWh in a battery).

Again we take the climate data from Palma de Majorca.

	Jan.	Feb.	Mar.	April	May	June	July	Aug.	Sep.	Oct.	Nov.	Dec.
kW/m^2	3,73	4,53	5,19	5,81	5,89	6,35	6,72	6,54	5,82	4,55	3,21	3,30

November is the month with the lowest irradiation, with a mere 3.21 kWh per day per square meter solar panel. We now do the calculation on the assumption that during the day, even with bad weather, we can collect sufficient energy to run the fridge.

At the end of the day, before a period of bad weather, the batteries will be fully charged; then comes the first night, then a whole day with bad weather, then the second night and then good weather comes once more. So in the worst situation we will need the battery to take over for two nights. As mentioned previously, we can power the fridge directly from the panels during the 10 hours of day light and we'll assume that the night is 14 hours long. We will neglect the fact that it will be cooler at night.

The fridge consumes 0.5 kWh per day, i.e. an average of 20.8 Watt. During the 10 hours of day time we will need about 200 Wh and for the 14 hours at night about 300 Wh. Since we need to store enough energy during the day to bridge two nights, we need enough solar panels to be able to collect 800 Wh per day.

Well, now we have to do the calculation backwards. If we assume that we have an inverter with an efficiency of 90% then we will need a supply of 220 Wh for the day and 330 Wh for the night. The battery has an efficiency of 80%; that means that the power needed for one night increases to nearly 400 Wh. We want to bridge two nights, so the battery needs to have an available capacity of 0.8 kWh. A battery with 100 Ah at 12 Volts has a capacity of 1.2 kWh so that we keep some reserve (always a good idea in a solar installation if it doesn't make things too expensive).

During a day, with 10 hours of sunshine, we have to collect the energy for the day (200 Wh) and the energy for two nights (0.8 kWh), which means that we need to collect 1 kWh per day. Now we come to the solar controller. We assume we have bought a good one with an efficiency of 95%, so its input must be 1.05 kWh.

In November we have an irradiation of 3.21 kWh per day per square meter solar panel. These panels have an efficiency of 15% so that they will deliver 0.48 kWh per day per square meter. For an output of 1.05 kWh we need 2.18 square meters of panels. The panels Juan Portales bought were 0.6 m^2 each. Since we can't get 3.63 panels we will actually buy four, each costing €86 so we spend 344 Euros for the panels.

Most of the time the battery will be nearly fully charged, so we can expect a life time of 10 years. Since we do the standard calculation over a period of 20 years, we will need 2 batteries. Lets have a look how much the **solar** driven **fridge** will cost us.

Fridge	150 €
4 Panels	344 €
Solar Controller	80 €
Inverter 1.5 kW	150 €
2 Batteries	200 €
Total	924 €

Our cost calculation will be:

Back payment per day	12.6 cents
Interest per day	6.3 cents
Payment per day	18.9 cents

Now we take the data for a gas driven fridge. Such a **fridge** costs at least 500 Euro to buy and costs of 36 cent per day for the **gas**.

Back payment per day	6.8 Cent
Interest per day	3.4 Cent
Gas per day	36.0 Cent
Payment per day	46.2 Cent

Our investment for a solar driven fridge was about twice as high as for the gas fridge but our costs per day are 60% lower.

I left out important facts? Well, you are right. I left out the costs for cables, connectors, for setting up the installation. But I left out as well that you would have to drive every week about 10 kilometers in order to buy the gas bottles not to talk about the time needed (strictly speaking you are not allowed to transport gas bottles in a normal car but many people in Spain just do because everything else is too difficult).

For the sake of fairness we have to compare these costs with a **fridge** connected to the **mains**. The fridge costs €150 and there are no other investment costs, so we pay:

Back payment per day	2.0 Cent
Interest per day	1.0 Cent
Mains per day	11.0 Cent
Payment per day	14.0 Cent

In order to get the complete overview we have a look at a **12V camping fridge** (the €500 are a bit optimistic).

Fridge	500 €
Panels	344 €
Solar-Controller	80 €
2 Batteries	200 €
total	1,124 €

Back payment per day	15.4 Cent
Interest per day	7.7 Cent
Payment per day	23.1 Cent

I haven't finished yet! We are going to look at a normal 230 Volt fridge which is **solar** powered during the day and **switched** over to **mains** during the night. When the compressor is actually working then the fridge will consume 75 Watts. The inverter has an efficiency of 90% so we need on the DC side 83 Watts. The solar controller has an efficiency of 95% so we need to get 87 Watts from the panels.

The panels have an efficiency of 15% so we need an irradiation of 580 Watts. One square meter of panel delivers 321 Watt (3.21 kWh / 10 h); in total we need 1.8 sq.m. or three panels of the type Juan Portales bought.

When the compressor is actually not running then we don't need energy. The panels are metaphorically twiddling thumbs.

Fridge	150 €
Panels	258 €
Solar-Controller	80 €
Inverter 1.5 kW	150 €
2 Batteries	100 €
automatic switch over	40 €
total	628 €

Since the battery is only needed to level out clouds passing by it can be much smaller and additionally we have the automatic switch (when the voltage of the battery drops below a threshold then we swap over to mains and the inverter is switched off)

Back payment per day	10.6 Cent
Interest per day	5.3 Cent
Mains at night	6.4 Cent
Payment per day	22.3 Cent

This relatively high price is caused by the fact that the panels cost money all the time but are used relatively seldom. We had calculated that the panels need to deliver 87 Watt. That is within 10 hours 870 Wh but the fridge only needs about (0.5 kWh / 24) * 10 = 208.3 Wh during the day time. The panels need to be four times as big in order to be able to power the fridge at any given moment.

We have a look whether a **buffer battery** is a good idea. For the sake of simplicity I leave out the losses. In our last attempt we had a factor of four because the compressor is running only 25% of the time. The idea now is that when the compressor is running then 25% of the energy comes directly from the panel and 75% comes from the battery; when it is not running the battery can be refilled completely.

During the 10 hours of day time the fridge will consume 200 Wh. Of this, 50 Wh will come directly from the panel and 150 Wh have to make a detour through the battery which has an efficiency of 80%. So 180 Wh have to go into the battery. During ten hours we have to generate 230 Wh so the panel has to generate at least 23 Watt. A 0.6 m^2 panel of the type bought by Juan Portales would generate 29 Watts in November (3.21 kWh irradiation per day makes 321 Wh per hour sunshine, with an efficiency of 15% that comes to 48 W/m^2; the size is 0.6 m^2 so we come to 28,89 Watt) and this would be sufficient.

Now we have to check the life time of the battery since it is used steadily. We take our standard battery with 100 Ah which will be good for 1,000 cycles. During its life time 1,200 kWh can flow through it. During the day time the fridge needs 200 Wh and 75% of this energy has to go through the battery which is 150 Wh. 1,200 kWh / 150 Wh = 8,000 days or 21.9 years. Theoretically the battery should last long enough (although I do have my doubts whether it would do quite as well in actual use).

Fridge	150 €
Panel	64 €
Solar Controller	80 €
Inverter 1.5 kW	150 €
Battery	100 €
Autom. switch	40 €
total	584 €

We have the cost per day:

Back payment	8.0 Cent
Interest per day	4.0 Cent
Electr. for the night	6.4 Cent
Payment per day	18.4 Cent

And here the ranking order of our different solutions:

Rank		Costs per day
1.	Normal fridge connected to mains	14.0 Cent
2.	Normal fridge, solar + buffer battery	18.4 Cent
3.	Normal fridge, solar; no battery	18.9 Cent
4.	Normal fridge, solar + mains	22.3 Cent
5.	Camping fridge, solar	23.1 Cent
6.	Camping fridge, gas	46.2 Cent

Conclusion: solar often pays for itself but not always! Admittedly I was a bit unfair using the irradiation data of November. If you use the fridge only during summer time the numbers are much better. You now know how to do the necessary calculations; maybe you use a spreadsheet program for correct results.

16.4 Air conditioning and solar power

If you look at a map of the world that is color coded to show the average irradiation over the year you will also be looking at a map of the world where air conditioning is not so much a luxury as an aid to being able to work during the day and rest at night. It's not just a matter of keeping the temperature at a reasonable level, but also being able to lower the humidity. Any air conditioning unit will do both; it makes life very much more pleasant and work that much easier.

Such comfort and efficiency comes at a price. Not just the cost of the air conditioner, which may be a mere €350 or so for a 2.5 kW unit but mainly the cost of the electricity required to run that machine (if such a unit is working day and night there will be €770 added to the electricity bill at the end of the year at 22 Cent per kWh). In fact, so much electricity is required to run that machine that if everybody in the street - and what's more relevant, everybody in the city - runs such an air conditioner, the power plant often cannot cope. Brown-outs (voltage lower than acceptable) and black-outs (no voltage at all) have become familiar to the average citizen living in a hot humid climate in modern and supposedly civilized countries. As more and more people all over the world look for the comfort and efficiency of air conditioning, the real cost to the planet as a whole will possibly become excessive. Could solar power possibly help?

Before we go any further, we have to have a look at how much work an AC unit has to do, and this in turn depends firstly on how hot (and humid) it is outside, and how cool (and dry) we want it to be inside. The greater the difference, the harder such a unit has to work. Well, that's pretty obvious, and there are all sorts of complicated calculations you (or a professional heating and ventilating engineer) can make to show that if you keep the difference between inside and outside temperature and humidity small, you use less electricity. On the other hand, people do want a certain amount of comfort - it's a question of balancing comfort against cost.

The second factor is how well your house is insulated. If you live in a country like Germany, Holland or Norway, your house will be extremely well insulated. True, it's mainly insulated to stop you feeling cold in winter, but that same insulation will stop heat coming in during the summer.

A house with foam-filled cavity walls built of thermal blocks will let a lot less heat in than one built of a single wall using no-fines concrete blocks. A solid concrete floor laid over 10 cm of expanded polystyrene foam holds back the cold much better than a hollow pot floor with just a cement screed cover (Spanish style). Double glazed windows with a good seal all round, proper awnings outside to shade the windows, a lobby between any outside door and the room being cooled, a roof with 20 cm of expanded foam under the tiles, all could help.

Building standards in other parts of Europe and most of the rest of the world can be a lot lower. In Germany the norm is to build low-energy-houses and there are even zero energy houses (even in a very cold winter night they don't need regular heating). However, there are other examples as well.

So the foreigner on one of Spain's Costas buys one of those nice white chalets with the fancy Spanish tiling and arched windows. The trouble with those chalets is that they are rather badly built, though you can't see that from the outside. Most of these chalets are built with "no-fines" concrete blocks, no cavity wall, no or very little roof insulation, no floor insulation, badly fitting windows and doors - everything leaks when it rains and you freeze in winter and suffocate in summer. So if he can afford it, the owner buys air conditioning to warm the house in winter but mainly to cool the house in the summer. Nothing to it, and why should they suffer when it gets really hot and the air gets really humid. Switch on the air conditioner and enjoy the comfort. In general, people with badly built houses can usually afford air conditioner units but not afford having their house rebuilt properly.

On first sight it seems to make a lot of sense to power an air conditioner unit with solar energy since demand and availability of energy fit well together. However, in the morning there are some hours when the solar energy is available but it is not hot yet and in the evening it stays hot even a couple of hours after sunset.

The two major methods that solar panels can be used to provide air conditioning are with or without the aid of mains. We are now concentrating on those installations in houses with mains connection (about air conditioning without mains see the supplementary chapter).

We buy a 2.5 kW air conditioner, a size which seems to be available everywhere in the world. There are two units in the cardboard box; one is a rather plain big metal box and the other one is a good looking (mostly white) unit for inside the house. In order to connect the two units we knock a hole in the wall. Now there are two copper tubes and a cable which connect the outer unit with the inner unit. In order to bend the tubes a wine bottle can be very helpful; if you kink the tubes you will have to call for a technician. When you have the tubes in a form that they fit to both connections you can do something which one should not do: you remove the stoppers and screw on the tubes. You should not take more than a minute for each tube. If you needed more, call the technician. Then you plug the cable into place and the big moment comes. You take the remote control and . . . the compressor starts to whir and the indoor unit starts to whisper. And some moments later cold air streams out. Bliss.

It says 2.5 kW on the label of the outdoor unit and most people are inclined to think that this means that it uses 2.5 kWh every hour, just like a 60 W light bulb uses 60 Wh every hour. In Spain that would mean that we would have to pay €5.50 per day for electricity. Or about 1,000 Euros per season (10 hours a day during 180 days).

Well, things aren't quite as bad as that. Maybe you remember the Coefficient Of Performance (CoP). On the box it states that the unit is rated at 2.5 kW and most probably you will find some numbers telling you how many BTU's it yields (that is British Thermal Units and isn't it really time that this measure stops being used, 40 years after the Brits stopped calculating that one pound is 20 shillings and a shilling is 12 pence). Ok, because of this CoP it will only use on average 400 Watts electric energy because each electrical energy unit will move about 6 thermal energy units (the CoP of an air conditioning is a lot better than the CoP of fridges because the difference of temperatures is smaller).

So you might be inclined to think that 4 standard solar panels of 120 Watts each might do the trick, plus a small inverter and everything works. Well, it will not work!

The first reason is that those 400 Watts are an average. When the compressor is actually running the unit will need 800 Watts (most air conditioning units have duty cycles of 50% when they are running full power; I suspect that this is in order to give the motor time enough to cool down in between). So we might think of a battery to level out the demands of the compressor.

How to get a realistic figure? We look at PVWatts and select for example again Palma de Majorca (the capital of the Balearic Islands) and find out that we can expect about 6 kWh irradiation per day per square meter panel in summer time. We use standard panels with 15% efficiency so that they will deliver 0.9 kWh per day per square meter; during the 8 hours of full sunshine one square meter delivers 110 Watts. The air conditioner consumes 400 W. We add a reserve of 25% and buy 5 square meters of panel for about 500 Euros. This should be sufficient if we neglect for the moment the losses of the other components.

Then we remember that electric machines need much more power when starting up and select a 5 kW inverter and hope that it will be strong enough (maybe we are able to make a deal with the shop selling the inverter that we can try it out for a day and test it). So that is another 500 Euros.

Now we come to the battery. When the compressor works the air conditioning will use 800 Watts but almost nothing (the little ventilator does not really count) the rest of the time. The solar panels deliver a steady flow of 400 Watts. When the compressor works it gets 400 Watts from the panels and 400 Watts from the battery. While the compressor idles we recharge the battery with 400 Watts and the battery will be full again next time. We select a 100 Ah battery for that job and get the following calculation:

Solar panels	500 €
Inverter	500 €
Battery	100 €
Work and small parts	200 €
Total	1,300 €

Again we work with a loan running 20 years and with 5% interest rate.

Interest rate per day	8.9 cents
Back payment per day	17.8 cents
to pay per day	26.7 cents

The air conditioning runs 8 hours with 400 Watt; that makes 3.2 kWh consumption per day. In Spain each kWh costs about 22 cents. With our solar equipment we only pay 8.3 cents. Energy costs cut down by 60%. Great!

But I was cheating. We left out the wear and tear of the batteries. Let's assume we bought a battery which can do 1,000 full cycles. Its capacity is 100 Ah at 12 Volts which means it can store 1.2 kWh. 1,000 cycles with 1.2 kWh makes 1,200 kWh through-put over the lifetime of the battery. We consume 3.2 kWh per day and half of it goes into the battery and out again (I neglect the losses in the battery). 1,200 kWh / 1.6 kWh = 750 days that is, the battery is used up in 2 years. We have to add to our calculation 9 Batteries more.

Solar panels	500 €
Inverter	500 €
10 Batteries	1.000 €
Work and small parts	200 €
Total	2,200 €

Again we work with a loan running 20 years and with 5% interest rate.

Interest rate per day	15.0 cents
Back payment per day	30.1 cents
to pay per day	45.1 cents

3.2 kWh costing 45.1 cents make a price of 14.1 cents per kWh. We do cut down our electricity bill, but only by 40%. Is there any possibility of leaving out the batteries? Well, not completely.

We used the batteries to level out the consumption of the compressor. We could just as easily install twice as many panels; when the compressor is not running the panels simply don't deliver electric energy (or somebody invents an intelligent circuit which observes the compressor and uses the surplus energy from the panels for something different like making hot water). But we do need a small battery for coping with the start-up current of the compressor; since the battery is in perfect condition (i.e. fully charged) most of the time it might last 20 years.

Solar panels	1,000 €
Inverter	500 €
Battery	50 €
Work and small parts	200 €
Total	1.750 €

Again we work with a loan running 20 years and with 5% interest rate.

Interest rate per day	11.9 cents
Back payment per day	23.9 cents
to pay per day	35.8 cents

3.2 kWh for 35.8 cents means a cost of 11.2 cents per kWh. We can run our air conditioning for half the price the neighbor pays. Perfect!

Well, now I will become a real spoilsports, because I would like to point out that we only use the air conditioner in summer time. Half the year it's just costing money and not doing anything (near to the equator this looks completely different). Effectively we have no financial advantage over mains! In fact it is not quite so bad as that because such an air conditioning unit can heat as well. In winter time it will probably not reach 2.5 kW heating power but even 2.0 kW or less will be very welcome to the inhabitants of a house (roughly comparable to a small electric heater with a noisy fan). If we don't need the AC unit for heating: there is still some use for free energy all the time!

But we still have to cover one topic. During the morning we might not need the air conditioning unit because it is not yet very hot and in the early evening it is still hot but we don't get enough solar energy any more. Well, if you can't make use of the solar energy in the morning, it is simply lost. And for the evening you need a little device which you can't find on the market (at least not in the way that you actually need): a mains changeover switch.

What does such a switch do? It simply disconnects your air conditioning unit from the inverter and connects it to the mains in the evening and the other way round in the morning (in case you had it running all night); then the connection to mains is interrupted and the solar panels take over again. But these switches are either quite expensive (about 500 Euros) or not quite legal. However, the biggest problem is that there is not a single device on the market which can detect the perfect moment for switching (if you know better, please let me know; that it is possible to build one I know).

There is one technical option left which we did not look at yet. In our last attempt we installed enough panels to collect the energetic needs of the air conditioning during the sunshine period and then swapped over to mains. Instead we could use batteries just for some hours in the evening. The batteries and the inverter are already there and we only need to enlarge the battery capacity. The panels were idling 50% of the time so there is more than enough energy to charge the battery during the day.

Normally one will need the air conditioning during 3 to 4 hours during late afternoon and early evening. During 4 hours it will consume 1.6 kWh. If we take two 100 Ah batteries they have together 2.4 kWh capacity of which we should only use 1.92 kWh. That gives us a safety margin of 20% which is nice.

If the battery can do 1,000 deep cycles with 1.92 kWh then the possible energy life cycle throughput is 1,200 kWh. Since we use every day 1.6 kWh it takes us 750 days to flatten the battery which equal 2 years. Since we use the batteries only during summer for 180 days the batteries will last 4 years. So we will need 5 sets of batteries.

Solar panels	1,000 €
Inverter	500 €
Battery	1,000 €
Work and small parts	200 €
Total	2.700 €

Again we work with a loan running 20 years and with 5% interest rate.

Interest rate per day	18.5 cents
Back payment per day	37.0 cents
to pay per day	55.5 cents

3.2 kWh + 1.6 kWh = 4.8 kWh for 55.5 cents means a cost of 11.6 cents per kWh. The kWh costs 0.4 cents more if we get the luxury of cool evenings. That seems to be affordable.

What would be necessary to make solar air conditioning cheaper? The answer is that the manufacturers of these units have to understand that there might be a huge demand for air conditioning units suitable for solar power. At present their equipment is designed and made to work well with the grid which can easily withstand high currents when the compressor first starts.

Let's have a look at a unit suitable for solar power. We kick out the 230V AC compressor which is now in the air conditioning unit and replace it with a DC driven compressor. The motor of the compressor must be able to run continuously and consume 400 Watts if available (else it slows down a bit). 400 Watts at 12 Volt would produce a current of over 33 Amps, which would be a bit too much (transistors able to cope with such high currents are expensive). For the 400 Watts we would need 4 panels; if we connect the panels in series we would get 48 Volts and a current of 8.3 Amps. That sounds better.

Now we would have the panels, a simple MPPT regulator and connected directly to the air conditioning unit. Now we get the following calculation:

4 panels, 120Watt each	480 €
MPPT regulator	120 €
Battery	50 €
Work and small parts	200 €
higher price compressor	50 €
Total	900 €

Again we work with a loan running 20 years and with 5% interest rate.

Interest rate per day	6.1 cents
Back payment per day	12.3 cents
to pay per day	18.4 cents

With a daily consumption of 3.2 kWh we come to a price of 5.75 cents per kWh compared to 22 cents from mains. We could reduce our electricity bill by 75%.

My message to the manufacturers of air conditioners is simple: if you let people know that you have something like this in stock, you could swap over to a three-shift-system in your production plants and you will still not be able to satisfy the demand! One billion people are waiting for such a product! What else do you need to get into gear? Even if you double the price at the same performance you could have more customers than you ever dreamt of.

17. Circuit breakers and faulty currents

We have already had a look at normal fuses. But there are other types of devices that do the same job (and similar jobs) just as well and these have the advantage that you don't have to dump them in the bin after they have done their work. This type of device is called a circuit breaker and there are so many types that one could write a complete book about them as well. So again we have to limit ourselves to describing what is really necessary.

All circuit breakers have an actuator and some kind of mechanism to disconnect the contacts. We have already seen that there are fast acting fuses and time delay fuses. The same applies to circuit breakers. A time delay is built in using a small hydraulic system (exactly how that works is of no importance here; but it is important to know that something like that exists).

The basic type of these circuit breakers is the line safety switch. As the name indicates the purpose of this switch is to protect the cable and not the appliances connected to the cable (these should have their own fuses or safety switches). If the fuse fitted to the appliance doesn't do its job (or maybe there is no fuse at all) then the cable might be in danger, in which case the circuit breaker takes over.

There are different circuit breakers for different maximum currents so they must be selected to reflect the cross section of the cable (and maybe how and where the cable has been installed). However, it is completely 'legal' to use a circuit breaker with a lower tripping current. In such a case the cable protection is really acting as an appliance protector. For any electrical installation within a house there should be a complete documentation stating which cables connect which sockets or appliances as well as the cross section of the cables. Without such a documentation it would not be 'legal' to increase the permitted amperage of that circuit again (because you don't know the cross section and the way the cables were installed).

It is very common to have a hierarchy of circuit breakers. For example, you have a main circuit breaker for the whole house (or at least one for each phase) and then a circuit breaker with a lower tripping current for each independent circuit (for example different circuits for sockets and light). The main circuit breaker should have a bigger delay built in than the other breakers in order to give the breaker guarding just the faulty circuit time to trip first (after all, there is no good reason why the TV should not work if the lighting circuit trips or vice versa).

Of special interest to us is a type of circuit breaker called an RCD in English speaking countries, which is the abbreviation of a "**R**esidual **C**urrent protection **D**evice". There is no reason why you should feel ashamed that this name doesn't mean anything to you as yet (in German this device is called either RCD or FI-switch where F stands for 'fault' and 'I' stands for a current, which doesn't make much sense either).

So what the hell is a 'residual current' or a 'fault(y) current'? Well, back to technical basics (and the explanation is a bit longer since you should understand why certain things are done in electrical installations). Lets assume that you have installed an isolated PV system with an inverter, and that you did all the mains (230 volt AC) wiring yourself using just two strands (that is all you need to get everything running); there is no potential equalization and no earth wire (what that is you will learn in a moment). If you don't connect any point of your circuitry to ground it has, as technicians say, free floating potential. Lets say we have some voltage sources with 100, 500 and 1,000 Volts outside the house. When we measure the voltage between 100 Volts and the installation we measure 0 Volts. When we measure the voltage between 500 Volts and the installation we measure 0 Volts. And the same applies to 1,000 Volts. So the potential of the installation adapts to any given reference (so it is floating freely).

That means that you can touch bare metal at any one point of the electrical installation without any risk. Touching two different points is something completely different; it might be deadly! In such a two-wire installation you wouldn't have any safety measures to protect you against faults. For instance, if inside your washing machine one of the cables touches the metal case (it shouldn't, of course, but this happens to be a machine made on a Monday) then if you touch just one of the two cables (trusting my statement that touching just one of the cables is perfectly safe) while leaning on the washing machine for support it is a case of "Bye-bye mate - I told you not to touch **two** points". So you had better plan on having some protection in your house; I will now tell you how.

Just in order to distinguish the two cables I'm going to call them L1 and L2 (don't confuse those labels with a rotating current!). Accepted practice for any electrical installation is that it should be impossible to touch any conducting part of any circuit (i.e. no direct access to L1 or L2) unless you use tools to get access (if you use a screwdriver to open the fuse box and touch the cables in there it is your own fault). I guess up to this point everything should be perfectly clear.

Now, so that readers in other countries can understand what I am talking about, a short explanation about sockets and plugs in mains circuits. Around the world 18 different types of plugs and sockets are used (says the German Wikipedia) and 10 of them allow you to push the plug into the socket in either of two ways. As you will soon see, there is no advantage whatsoever for those systems where there is only one possibility. For my explanation I use the German system where there are two possible ways of pushing the plug into the socket.

Now we put some appliances into the house. First we have the old toaster inherited from Aunt Carola; a very nice design and really easy to clean since most of the outside is stainless steel. Over the years the cables inside of the toaster have become a little frail and porous (but you can't see that from the outside). One fine day the toaster dropped from the table and now the bare L1 cable touches the casing (Fault no. 1).

You also got a beautiful standard lamp, made of brass, from Aunt Carola. Uncle Rudi repaired it long ago, long before you got the lamp, but he made a botch of it; instead of connecting the yellow-green wire (he just cut it) to the body of the lamp, he used one of the other strands (Fault no. 2). Purely by chance, all the years the plug had been pushed into the socket in such a way that there was no danger. So we have two faults in our system but actually there is no immediate danger. Not yet!

One fine day you want to bake some cakes and the toaster gets in the way. You unplug it and put it to one side. After finishing the cakes you put the toaster back in place and, purely by chance, you plug it in again, but this time with the plug pushed in 'upside down'. Now we have a deadly danger in our household. Between the lamp and the toaster we have 230 Volts AC. This can go well for years or for just a few seconds -

just as long as you don't happen to touch both the toaster and the lamp at the same time (even if one of them or both are not switched on).

In order to prevent such situations it is mandatory that there must be a potential equalization (what this is I explain right now). The bare metal surface of any electric appliance has to be connected to this potential equalization. In this sense even your water installation is an electric appliance if it uses an electric pump. The same applies to your central heating (pump!) and the under structure of your PV installation. So everything made of metal that could, due to a failure, have contact to one of the strands of mains must have a contact to the potential equalization. This includes all the protective contacts of all sockets (often called 'earth contacts' which is correct if you have a mains connection) by help of the famous yellow-green cable (normally 2.5 mm^2 cable will do; if in doubt it has to be as thick as the thickest wire in the circuit). If you have a central heating with many radiators and all the tubes are metal then it is sufficient to establish a single contact. The complete conductive interconnection is called potential equalization because all these metal parts are on the same potential (what potential it is does not matter at the moment; we only make sure that it is the same).

Now we have a look at the toaster. The metal case of the toaster is connected to the potential equalization and either L1 or L2 are in contact with the case. That is no problem because the potential of that cable is pulled to the same potential as all the metal installations in the house have. You could say that this is a generalization of the principle: it is always save to touch a single point in a potential free electric installation.

We now take L1 and connect it at a central point to the potential equalization. Now the complete installation within the house has a free floating potential in reference to the surroundings but the potential of L2 is not floating free in reference to the house. The effect is that if I put in the plug of the toaster and the case is connected to L1 then nothing is going to happen because is has the same potential as the potential equalization anyhow. If I put in the plug the other way round there will be a short between L2 and the equalization which in turn is connected to L1 and the fuse will blow. The fault in the toaster can't provoke any danger any more.

With the standard lamp the situation is different because the yellow-green cable was not connected to the metal. Depending on which way round the plug goes into the socket it is either harmless or it offers a lethal effect. Conditions like this are the reason why many electricians say that laypersons must not alter (or try to repair) electrical installations and they are right (they did not bribe me; they are simply right). If you think that they say this is just in order to protect their sinecure you might be right as well; but most people have no idea about how many Rudis are running around in the world!

By introducing a potential equalization network we can't be certain that there are no faults within our installation, but we can make sure that these faults will cause no harm (unless Uncle Rudi tinkered with the cables; by the way, you can use a multimeter and measure the resistance between the case and the protective contact; if the resistance is zero there is definitely no dangerous fault; if you measure something different there might be one).

Generally nobody knows the term 'potential equalization' and if the term is known it is confused with earthing; reason is that in a normal house installation (when there is a connection to mains) they are identical but we are talking about solar installations and therefore we have to distinguish the two. At the moment you connect the potential equalization with ground (by help of an earth anchor or using the GND cable from mains) you have an earthing and the cables in the house get new names (the potential equalization with the yellow-green cable is GND or protective ground; the cable connected to GND is now called N for neutral and the last cable is called P for phase). When you have a mains connection the earthing is mandatory and when you have an electric island then it is an option (we come back to that in the next chapter) but it does not change anything in respect to safety.

In order to finish this subtopic (the real topic are circuit breakers) I shortly explain why earthing is mandatory if there is a connection to mains. In the distribution grids they use just three cables; normally a fourth one would be necessary for the compensation current in case that the load on the three cables is not symmetrical. These currents are normally relatively small so the owner of the grid uses the Earth as cable number 4 because it's much cheaper. When the cables come to your house N is already identical with earth. So the compulsory earth anchor in the house makes sure that your installation follows the standard.

This was, once again, a long introduction. But it was necessary as a basis for understanding what purpose is served by the RCD (residual current device). We just saw that with the way we wired our system we were able to make sure that no dangerous situations could occur. The circuit breaker will disconnect the consuming devices immediately whenever a dangerous situation arises. OK, let's assume that we have a 16 Amp circuit breaker for all the devices in the kitchen. A joint in the supply pipe to the water tap is leaking. Drop by drop the water is falling down on to the cable connecting the fridge to the socket. The drops run down the cable and into the socket. The water makes contact with the L2 strand and then runs down the wall to the floor tiles. Here it moistens the mortar between the tiles and makes it conductive.

It is a brilliant sunny morning and you look forward to your breakfast. You walk with your bare feet into the kitchen in order to boil some water for a nice pot of tea. Your feet are (partly) on the moist grouting between the tiles and you touch the tap (a very pretty and quite expensive solid brass tap). Now we have a current that runs from the L2 strand in the socket, via the floor and your body to the tap and further to the L1 strand. Your muscles cramp (so you can't let go of the tap) and your breathing stops as well. The circuit breaker muses a little and decides: only 50 Milliamps so everything is perfectly all right, we are still in the green area (the human body has a resistance of approximately 5 kΩ; 230 Volt / 5 kΩ = 46 mA)! What you should know is that starting with 10 milliamps cramps can develop and with currents between 30 to 50 milliamps the breathing might stop; a normal circuit breaker has no chance to sense such a small current.

The RCD circuit breaker counts electrons coming in through strand L2 and going out through strand L1 (and vice versa when the current in our AC circuit reverses itself). As long as the number of electrons in these two strands balance, the RCD circuit breaker is "ON" and allows

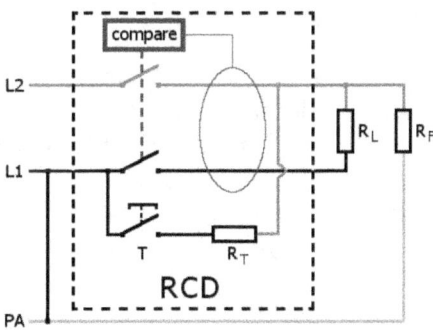

current to flow to appliances such as R_L via strand L2 and back again via strand L1.

However, when you stand on the damp kitchen floor, the electrons flow from strand L2 through your body (R_F) to the potential equalization and from there to L1 but before the counter. There are more electrons coming in through L2 than are going out through L1 (as far as the RCD is concerned) and the RCD trips and disconnects the circuit. What's more, it does it so fast that no lasting damage is done to the human body.

In the picture you can also see a push button (T for test) and a resistor R_T; this resistor has the same resistance as a typical human body. When the button is pressed the counter seems to think that somebody is in mortal danger and the RCD will trip immediately. If you have worries about being electrocuted, press that button two three times per year to check whether the RCD is still on guard.

So now we have the explanation for the names of this device. Sometimes it is called an FCCB (Faulty Current Circuit Breaker) because it breaks the circuit whenever a fault indicates that there is a current running along an unintended path. Other people call it an RCD (Residual Current protective Device) because if, when comparing the current on L1 with the current on L2, there is a difference, then there is a fault and so it is essential to cut the circuit.

Normal circuit breakers only switch off whenever the current is too high, i.e. whenever there is a short or whenever the connected appliances consume too much power. However, they simply cannot detect a leakage current. A RCD (or FCCB) can ONLY detect this kind of current (whether a current of 5, 10 or 30 amps is running through L1 and L2 does not matter to it at all).

Actually, there are no counters in the circuit breaker (and Father Christmas doesn't exist either). The RCD uses the effect that any current produces a magnetic field; this switch is designed in such a way that the magnetic field of L1 annihilates the magnetic field of L2. Only when the currents are not equal a magnetic field is detectable from outside and then the switch is tripped. In fact, RCDs can be extremely sensitive.

We come back to the little chalet in the Andalusian Mountains. Jan de Witt had decided to invest in a photovoltaic installation and he felt sure that he could do nearly all the work himself (most of the switches and sockets in the house were not where he wanted them to be and he saw no problem in cutting some slots in the walls with an angle grinder, in which to put cables and plaster over them later on). First he drew a plan where he wanted the switches, sockets and lamps to be. Then he decided where the batteries would go (a little extension on the north side of the house) and the electronics. The fuse box could stay where it was. He ended up with five circuits (each with its own circuit breaker) and a single RCD.

He then ordered all the components and started to work (since he had no electricity yet he rented a generator and some tools from a hire company). He cut a couple of slots into the walls and placed the strands in them (he had seen a nice trick Spanish builders used; they put the cables in and fix them every 30 to 40 Centimeters with a handful of plaster of Paris; the next day the slot can be plastered without the cables falling out all the time).

At the beginning of March all the components arrived, one by one, and he set up his installation. After a rainy weekend (yes, these do happen, even in Andalusia) everything was ready; the battery was fully charged and the inverter worked well (he tested it with some light bulbs). Now came the big moment. He pushed the switch of the RCD up but then, when he pushed up the switch of the first circuit breaker, he heard a pronounced "Click". The RCD had switched off. He tried all the circuit breakers, one after the other, but had no luck at all.

Jan de Witt remembered that such a fault can be singled out only by help of the strategy which Julius Caesar laid down, the "Divide and Conquer" system. He unplugged all devices (in a mood of pleasant anticipation he had even connected his old stereo equipment). To no avail, the RCD refused to give any service. No matter what he did, there appeared to be not a single circuit without fault. Fighting back his tears he walked down into the valley to drown his sorrows in Paco's bar.

The next day there was marvelous sunny weather and the sun burned mightily. Jan looked at his photovoltaic installation as if it were a traitor (especially the fuse box). Late in the afternoon he got over his depression and started to check each and every cable, socket, connection, just about everything and anything. No result, everything seemed to be as he thought it should be. Early that evening he connected some normal extension cables to the inverter in order to have at least some light; this seemed to work, so it couldn't be the inverter.

Later that evening, with a glass of red wine in his hand, he once again stood in front of the fuse box and stared at it as if mind power could do something. Just in order to demonstrate that he had not yet completely given up, he tried the various circuit breakers once again . . . . and suddenly was perplexed. When he switched one of the circuit breakers on, the RCD did not say "Click!". A self-repairing fault? That can't be! But it happened!

Since you have so carefully followed my little tale I will give you the solution to this riddle. Everybody thinks, when hearing the word 'insulation', that this is some special substance which allows nothing to come through, but this view is wrong. Strictly speaking there is no such thing as insulation because something can come through any substance; the thing is that normally the extremely small amount that comes through the insulation around an electrical cable is not relevant in technical systems. Jan had placed the single strands directly in the outside wall and the rain moistened the walls (don't tell me that Spanish builders use water-resistant paint if not paid extra). So there was a small current from L2 (the phase) through the isolation into the wall and down to earth and via the earth electrode of the house and from there to L1. The current was strong enough to make the RCD suspect that a person was in danger. The RCD was just doing its job, and doing it well!

Now Jan de Witt knew why plastic tubes should first go into the wall and then cables into the tube. It took him another few days to open the walls once more and lay the cables again (just a tip: put the corrugated tubes in in such a way that no moisture can leak in and - just in case that water does leak in - that it can leave the tube again. That means, no corrugated tubes in the floor!).

After everything was working perfectly he phoned the electrician for the acceptance test (he did not really need one since there was no insurance for the house anyhow; but he wanted to be sure that he had done nothing wrong). The day before the electrician came he disconnected the cable between the inverter and the RCD and using an extension cable he connected the fridge and some lamps to the inverter.

Next afternoon the electrician came and had a look at the documentation and then started his inspection (Jan had left open all the junction boxes and sockets). When he came to the fuse box the electrician followed with his index finger every cable (Jan had put quite a lot of effort into this work; all the cables were just long enough so that one could push them aside in order to have a better look at other parts of the cable network and there were no cable clumps). It was only a few minutes later that the electrician took out the form, filled it in, signed and stamped it.

While the electrician drove off Jan went to the fridge and took out a perfectly chilled bottle of Cava (the Spanish equivalent to Champagne) of the 50 Euro class. With the bottle and a glass he went to his little terrace and set down in the early evening sun. He filled the glass and raised it for a toast to the electricity mast down in the valley. Jan was glowing with pride! It had not been easy but he was looking at the proof: even economists can successfully plan and set up a photovoltaic installation.

This chapter had not the intention to tell you how to connect 230 Volt AC installations; it is meant to give you the understanding why some things are done in a special way. Many amateurs leave out an RCD in their installation in order to save some money; they even leave out the potential equalization or earthing as well. I hope you understood now why that is no good idea.

If you make a circuitry with a 12 Volt battery then there is no danger (if you insert a fuse directly at the plus pole). So again: it is not for your fingers to connect dangerous voltages (higher than 50 Volt AC or 120 Volt DC) to any cables. If you do and somebody gets injured or even killed it is your full responsibility! This is the job of a fully qualified electrician and additionally your insurance company has no cheap way out.

18. Lightning protection

This is really a very complex topic. So once again I delimit myself to explaining the very basics. You must not take them as a basis for the planing of an installation (if you do it is fully on your own risk; a better solution is to ask somebody who got the qualification to do the technical acceptance test for lightning protection).

Thunderstorms and lightnings even without a thunderstorm are possible (nearly) everywhere and at any time. There are areas where the risk of lightning strikes is relatively high and there are places where this risk is nearly not existent. For example if your house stands directly beside a pylon of a high tension transmission line (if a lightning strikes it will go into that pylon but you can't be absolutely sure that there will never be an exception).

First we have to have a look what happens during a lightning. On one side we have an electrostatic charging of layers of air and on the other side we have the potential of the Earth. By help of collisions (most probably with cosmic radiation) every now and then an atom or a molecule gets ionized. Because of the tension between air layer and earth potential the ion and the electron move apart. Normally it is only a question of a very short time for each of them to find a new partner. But if the tension is that high that they can accelerate enough to ionize other atoms or molecules then we have the starting point of a lightning.

With an incredible pace a conductive 'filament' (actually experts think that it is as thin as a finger) gets longer and longer and finally it connects to the Earth. The more electric energy flows through this filament the hotter and wider it gets (it conductivity increases really fast). Between air layer and Earth one can have easily 1,000,000 Volts so a gigantic amount of electric energy is converted in a very short time into heat. In this plasma channel currents up to 100,000 Ampere are flowing and the air explodes.

Knowing this we can start to deal with the question where a lightning likes to strike. In physics classes there is a nice experiment. In front of a plane metal plate a midsize metal globe is positioned. Then a tension is connected between plate and globe and the tension is increased until there is a stroke. Then the experiment is repeated with a smaller globe and in the end a metal thorn is used. The result of the experiment is: the smaller the globe the lower the voltage to start a stroke. That means that a lightning likes to stroke into spiky conductive objects pointing to the sky (if you go for a swim during a thunderstorm that little bit of head poking out of the water might make quite a difference; so don't do that). Trees and buildings are favored targets as well.

So, back to photovoltaics. You have a couple of panels on the roof and from there some cables are going down into the house to the solar controller and the battery. You connected the minus pole of the battery with earth potential because something similar is done in cars. And you connected one of the cables on the AC side with earth as well. Now a thunderstorm passes by and the tension between earth and clouds rises rapidly. And now you might find out that you set up a lightning lure installation.

The leading lightning (this thin ionized air filament) dashes down in direction of Earth looking all the time for the easiest way to go (with the smallest resistance) which is why lightnings are zigzagging. Now the filament comes near to your house and senses that the resistance through the photovoltaic installation is electrically much shorter than through the air (somebody was so nice to place very well conducting cables). After 1,000 meters or more through the air one millimeter of glass or a thin layer of plastic insulation is no big deal.

Now we have to deal with 100,000 Ampere which want to crash through our PV installation. So be prepared that cables and sockets are going to shoot out of the walls and all electric consumers which by chance were switched on at that moment (this includes probably the expensive sine wave inverter as well) will be scrap; some devices not switched on are killed as well but they had a better chance. That lightnings inflame combustible materials is no made up horror story so the catastrophe can be enlarged easily. Lightnings are something you definitely don't want within your house!

Now we come to the second question. How can we prevent that a lightning nibbles off our solar equipment? Absolutely easy: offer something irresistible to the lightning (something which is much cheaper for us than an inverter).

At this point of the story I have to make clear again that if you do it wrong you might produce more damage than good (maybe without our offer the lightning would not have been interested in our installation at all or you lure it to the most expensive parts). Ask a specialist on this topic or all the risk is yours (suppose you build the lightning protection yourself, a lightning strikes and there is a considerable amount of damage; you phone the insurance company and they send their expert; you can be sure that he will find something done wrongly and you will get not a single dime from them; possibly there is even a third party damage and you will have to pay for that as well; just a warning).

When you have a look at a lightning protection on a normal roof then you will see right on the top of it a horizontal rod (will be seen by a lightning as an appetizer) and every few meters there is a relatively small rod sticking up (irresistible for a lightning). The correct name for it is arrester rod. Normally all these rods are made from aluminum (aluminum has some kind of self-protection against corrosion). All these rods are interconnected and there is at least one rod leading down at the outside (!!!) of the house and is connected there with the ground anchor of the house (with large roofs there can be quite a big number of these lightning drains). If a lightning sees that installation it will jump for it (in the true sense of the word).

So, the leading lightning comes dashing down and senses the conductive rods on the roof, especially the rods sticking up to the sky. That is because electrons are leaving the rod and move in the direction to the lightning (in seafaring this effect is called Saint Elmos fire; shortly before or during a thunderstorm there can be a slightly blue shimmer around metallic parts especially if they are pointing upwards; this light is produced by electrons shooting up and ionizing air molecules). This indicates to the lightning that there is an electric short cut.

I shortly return to fuses. There I explained that fuses can cope with pretty big currents for a very short period of time. If there is a high current for too long a period then the metal filament will melt. Nothing else is a lightning protection. The energy of lightnings is known (well enough) and the rods have a cross section big enough for getting the energy of the lightning safely down into the earth (before they melt). So ultimately this is what a lightning protection does: measured in meters the way around the house is much longer than the way through the house; but electrically seen the way through the rod is by far shorter and the lightning will take this detour. And that is exactly the aim of the complete exercise!

OK, we successfully got the lightning around the house; how does it continue? When the lightning comes to our earth anchor it will hit problems because earth is by far not as conductive as the ionized air or the aluminum rods. Though it is not the correct way to do it we assume that we use a single peg in order to get the lightning into the ground. The energy reaches the peg and looks again for the easiest way to go. The energy produces a cone with higher voltage and the cone expands in order to find an easy way for the energy to continue its way.

We already saw that a current flowing through a resistor produces a voltage over the resistor. Therefore the potential measured between the peg and earth potential in the distance shoots up since the current is enormous and the ground has a relatively high resistance. Converting its energy within this big ground-resistor into heat is the death of the lightning and everything is over.

The only thing in this sequence we don't like is the increase of the local earth potential. If we take some more pegs and interconnect them then each of them will start a coniformed earth-resistor. In this way we get a much bigger volume where to convert the electric energy into heat; or the other way round, the resistance is much smaller and therefore the peak value of the local earth potential is smaller as well.

Normally one will not use pegs but the complete foundation of the house as an earth anchor (the reinforcing bars will be in the concrete anyhow and they are all electrically interconnected; well, concrete is no good conductor but earth isn't either). These bars are there and we can use them for free as a really big earth anchor. If the house is relatively old there will possibly be no foundation of this kind. In that case we need to use pegs and distribute them over a relatively big area. Concerning the spaces between them: ask a qualified expert because it depends on too many factors!

Every now and then somebody comes up with the proposal to use an old engine hood as an earth anchor for the lightning protection because it has quite a big surface area. Out of two reasons this is no good proposal. First, electrically seen, the difference between a single peg and the hood is not so big (maybe an advantage for one millisecond). It is necessary to make the cone of energy flow much bigger. The second reason is that such a hood will be completely disintegrated by rust within a few years time. During the first few years there might be a very small advantage of the hood compared to a single peg. But then, without you knowing it, the lightning protection converted into a lightning attractor with nearly no protection (the expert of the insurance company will for sure find this botch)!

I did not give you all these explanations without a good reason. Without them you would not be able to understand the rest of this chapter. Lets assume we have a bog standard house installation with earthing. And where will the earthing be connected to? The foundation! And where is our lightning protection connected to? The foundation! If a lightning is discharging through our earth anchor then the potential of this anchor can be easily a few thousand Volts higher than the earth potential in the distance. Since the complete house does this voltage jump there is no danger for the inhabitants.

The problem is that for a few moments our local earth potential is much higher than the AC-voltage coming from outside. If we don't do anything about that effect then we lured lightning energy into the house. But you can see it as well in a pragmatic way: the house was saved and the biggest part of the destructive energy is gone elsewhere. So we come now to the question how we can deal with the rest.

The lightning protection has three layers (gross, medium and fine protection). On the first level (that is what we just had a look at) we caught and guided the lightning around the house and got it into the ground. In order to prevent that some of this energy is looking for an alternative way (for example cables in the house on the other side of the wall) there should be a distance of 50 cm or more between the rod coming down from the roof and other installations. With that we have the gross protection which is called (at least in Germany) 'type 1 protection'.

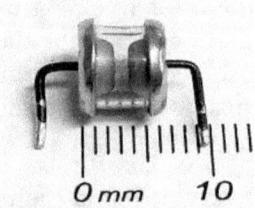

Without some additional protection the rest of the energy of the lightning would run back and forth in all the cables within the house. Here comes in the medium protection which should be installed at every cable entering the house (you might have guessed that this protection is of type 2). It is realized by help of over-voltage conductors. They are used to protect complete switching systems or houses. In the picture you see such a part (source: Wikipedia; Ulfbastel; public domain). The advantage of these parts is that they are relatively cheap (3-5 Euros per part), reliable and they have a very high energy conversion capability. The only real disadvantage is that after the flashing arc is ignited it will stay even with relatively small voltages. This can easily be compensated by help of a circuit breaker or a fuse being in line with the over-voltage conductor with a delayed characteristic.

The over-voltage conductor is placed between the AC cables and earth. Since the polarity of the voltage does not matter the power impulse with the high voltage will be converted into heat within the over-voltage conductor. If the voltage peak comes from the outside (that could be a telephone cable as well) the energy is discharged to earth potential. If the lightning stroke into our house then the local earth potential will be too high and the energy is drained to the AC cable. In any case, by help of the over-voltage conductor we reduced the amount of destructive energy by an order of magnitude. With the rest of the energy the devices within the house or in the AC distribution grid have to get on with.

On the next level (I think everybody now guessed that this is type 3) we have the protection of appliances (within the EU the CE label indicates that the device has such a protection; how good it works is a different question). Here normally protection diodes are used (I talked about them already in the chapter about electric machines). The amount of energy they can cope with is much smaller but on the other hand they react very faster.

I think it is obvious that type 3 protection is of no help in case of a lightning if there is no protection of type 1 and 2.

Meanwhile you will understand that lightning protection is indeed, as I told at the beginning of this chapter, a pretty complex topic. If you think (or know) that your house is in danger then you have to do something on all three level. Measures of type one make sure that no cables or sockets are rocketing through the living room but quite a lot of your appliances might be killed. Measures on level two make it possible that the appliances can cope with the rest of the energy. If you have protection on all three levels then the chances are pretty good that the complete household survives a lightning undamaged.

We are not finished yet because we still have to look at what will happen if a lightning strikes near the house (so it does not matter whether we got level 1 protection or not).

Lets assume we have a standard lightning with 100,000 Ampere. Any current produces a magnetic field and when a magnetic field is changing it induces a voltage into any conductor (and these voltages can be pretty high because the current in the lightning changes rapidly). We have a PV installation on the roof and the DC cable runs down the wall on the outside. This cable works like an aerial (as all the other cables in the house as well) and the lightning produces shock waves on these cables but our DC cables are probably the longest. There will be multiple reflections within all our wiring and it is unpredictable when and where the waves overlap and produce a really high voltage. So this energy is looking for victims and not opponents and very slowly the resistance of the copper weakens the shock waves (converting the electric energy into heat).

We will need two over-voltage conductors on the DC side and you will need two of them on the AC side as well. If there is a telephone cable then you will need there two more. Since you can't know whether the over voltage from the outside will be higher than the one inside of the house, you best place the over-voltage conductors where the cables enter the house (if the lightning stroke hundreds of meters away but near to the power line then the destructive energy travels over the AC line and then the best place is near to the connection point anyhow). The over-voltage conductors are placed between the cable to be protected and the local earth potential. If the potential difference is high enough a flashing arc will start (if you use the type shown above) and it will convert as much electric energy into heat as possible.

And there are people who claim that shielding is the best way of preventing such shock waves. That means that all the cables are within metallic tubes which are earthed. You could do that as an additional means if you really have severe problems with lightnings. But I think it is some kind of overkill which is only worth doing if you want to protect your house from a nuclear EMP (electromagnetic pulse) as well. A well dimensioned over-voltage conductor at any strand coming into the house should normally do.

Well, this was now five pages with basics of lightning protection and now we are coming to the PV installation at last.

We start with the case that the house already had a lightning protection. So now we come with our panels and the under construction for them, installing conductive elements on the roof. And even worse, we put cables which go into the house. There are ways of calculating the areas being protected by the lightning protection (an expert has to do that; one way is to place a thought globe with a diameter of 45 meters on the arrester rods and if the solar installation is not touched by this globe then it is in a save area). If the complete PV installation is within this save area then there is nothing to worry about (but you should place two over-voltage conductors where the DC cable goes into the house); the lightning can't sense our panels.

If the PV installation is not completely within the protected area then either the PV installation needs to be adapted to the lightning protection or vice versa. What is possible and what will come cheaper only an expert can know. Sorry, no better news.

If there is no lightning protection then we have to distinguish two cases:

> A) You have an isolated PV installation and not a single part of the electric installation is earthed.
>
> B) The protective conductor (the yellow-green cable) is earthed (whether you have a connection to mains or not makes no difference).

A) The complete PV installation in invisible for the leading lightning because electric-wise the installation doesn't bring the lightning any nearer to earth potential. But that does not hinder the lightning to strike somewhere near to your house or into the roof gutter. So you will need at the appropriate places over-voltage conductors. For running a PV installation it is not necessary to earth any part of it (neither DC nor AC). But what we have to do is to interconnect all conductive parts with the famous green and yellow cable (this interconnection is then called potential equalization and it is necessary to guarantee the safety of persons; more about this topic in the last chapter).

At the inverter there are two connectors on the AC side and between them you can measure 230 Volts AC. You can touch the copper within such a cable without any danger because the complete circuitry is potential-free. We could select any point we like within this circuit in order to earth the installation but it is not compulsory to do so (if you touch the copper of both cables this might have been the last experiment you ever did).

What you must do is interconnect by help of a cable all conductive parts that can be touched; this is called a potential equalization (this is not earthing; for an earthing you must connect the potential equalization with earth potential, for example with the earth anchor of the house). And the next thing to do is to connect one of the two AC cables with the equalization. Since the metallic under construction of the panels is (at least in Germany) a conductive and touchable part of an electric installation it needs to be connected to the equalization as well. For the lightning that makes no difference at all; it can not sense it anyhow.

B) If you connected the minus pole of the battery with earth then all the panels and all the cables on the roof look like juicy prey to the lightning. I can't see any reason why somebody should connect any pole of his battery with earth; but if you do, it is your obligation to present something even more delicious to the lightning; that is, you have to think about a lightning protection.

Well, next case. The battery is not earthed but your 230 Volt installation is. If there is no transformer in your inverter this behaves as if you earthed a pole of the battery; that is, you have to think about a lightning protection.

Next but one case: there is a transformer in your inverter so we got an isolation between the DC circuit and the AC circuit. Now comes the bad news. The electric strength of the transformer will be some thousand volts and the lightning is probing with a million Volts. If there is the smallest leakage current then some electrons will invite the lightning to have a look at the inside of a PV installation; that is, you have to think about a lightning protection.

If you have mains in your house then the equalization is connected to the earth anchor of the house (mandatory!) and it is then called protective earth. Since it is compulsory to connect the metallic under construction of the panels with protective earth then the PV installation might be a lightning attractor. If it will be depends on the local conditions. Lets say your house has a high and steep roof and the panels are near to the lower edge of the roof, then probably the risk is low. If you have a flat roof and the PV installation delivers the highest point of the building then the risk could be considerable. Anyhow, you will have to think about a lightning protection.

Upshot: if there is **no connection** whatsoever between your PV installation and **earth** potential then the **lightning is not interested** in your cables or your panels.

If you have, out of whatsoever reason, a direct or indirect connection to earth potential you should make up a risk analysis. How often are there thunderstorms and how expensive is your installation and all the appliances connected to it (when did you last time make a copy of the hard disc and how much will it cost to do all that work once again?). The opinion of an expert who knows the area might be very helpful.

If you really understood the complete chapter and there can't be any fights with an insurance company then you might decide to set up your own lightning protection system (something is better than nothing if properly done). How thick the rods or the cables have to be you can easily find out via Internet. Or even better, ask advice from a local expert.

The necessary materials for a lightning protection are relatively cheap; what makes it expensive is the necessary labor and the knowledge about what has to be done and where everything has to go (how many arrester rods and where and how to do the earth anchor if the house hasn't got one). Since it concerns not just your safety but the safety of other people as well you should in ANY CASE get a qualified technician to do the acceptance test and you should keep the protocol.

19. Guerrilla photovoltaics

Several months ago, what have become known as guerrilla installations, showed up on the market. These are complete packages with one or two solar panels, an inverter and all the small parts you need to fix the panel(s) to, for example, the balcony balustrade. The complete installation is connected to the house mains circuit with an ordinary plug which is simply pushed into a normal socket.

When the sun shines, this small installation feeds electric energy directly into the mains and can power the household appliances or even feeds into grid - your main electricity meter will run backwards (unless the meter has a reverse lock). The sellers state in their advertisements that these installations are absolutely legal and should pay back their costs within a few years.

Maybe you remember what I said about the Dutch solar shop. These people know their prices; they offer these guerrilla installations at a price such that you can just about save a little bit of money. These guerrilla marketeers buy this stuff for a ridiculous low price in Hong Kong or Korea and they make the profit, not you, but that's OK, it is what merchants do all over the world. Funnily enough they claim the same performance data in every part of Germany (and there are really quite big differences in the amount of irradiation reaching Hamburg, Düsseldorf or Munich). I think you can assume that they took the best irradiation data available for Germany and then calculated output assuming perfect orientation of the panels. It is not really likely that you will have such a perfect situation and perhaps they massaged their numbers just a little bit - perhaps.

Hopefully you will remember that AC voltages over 50 Volts can be fatal. Now you have a look at the plug and what do you see? Two blank metal pins. No problem, says the seller; integrated into the inverter is an electronic safety circuitry (and if he really wants to impress us he will mention that it works with sophisticated micro processor chips), and whenever this circuitry can't sense the right voltage with the right frequency on mains then it will disconnect immediately. The inverter will only work when the plug is fully pushed into the socket. Absolutely safe!

You buy such an installation, you install it and you are happy that you have to pay the electricity company just a little less in the future. A few days later your children are playing in the living room and your sun Max who is very inquisitive at all times („If I were not meant to touch anything and everything what are those little things at the end of my arms good for?") pulls out the plug from the socket and sure enough, the inverter stops delivering electric energy to the plug. Then Max plays racing cars by pushing and pulling the plug on the carpet of synthetic fibers. Finally he presses his 'racing car' into the neck of his little sister Nina.

Did you ever hear the term 'static charge'? If not, have a look at Google and Wikipedia. With static charge you can destroy any electronic circuit (unless it is especially protected; you find such protection mostly in military equipment) and rubbing on plastic is a good way of producing static charges. When a static charge discharges (through the electronic components) that is like a lightning in miniature. We can have three possible situations now. First is, that the solar installation works as before. Second is that inverter is destroyed. Third is that only the safety circuitry is destroyed and this is the dangerous one! In this case Max might have electrocuted his little sister (I don't think that the risk is very high but it is there!).

All electric installations are (or should be) based on a concept called 'one fault safety'. Electric installations must be carried out in such a way that any single fault can not cause a dangerous situation. The first fault, the static discharge to the safety circuit, converted the inverter into a potentially deadly weapon. This applies to all plug and socket combinations worldwide.

Within the EU there must be a declaration of conformity before any technical device may be sold. In this declaration the producer or the seller of the product states that the product conforms to all technical regulations within the EU and naming each and every one of the relevant regulations. After making this declaration the manufacturer or seller can file it (the declaration will neither be checked by bureaucrats nor by technicians) and he has to attach the CE label on the product. Without a CE label it is illegal to sell a technical product within the EU.

Since it is not reasonable to expect a normal client to be able to check whether a technical product is safe or not, his responsibility is very limited. At the most he can check that there is a CE label but normally he can expect that there must be one when he buys the product within the EU. The next in the responsibility queue is the seller. If there is no CE label on the product he will have an intense conversation with the custodial judge the same day. Next in the row will be the wholesale trader or the importer. Ultimately the one who brings a product onto the European market is the one who has to make sure that there is such a declaration and that the CE label is fixed on the product (with bigger devices a copy of the declaration has to accompany the device). The CE label is a registered trade mark of the EU all over the world. If nothing else helps the manufacturer will face a charge because of trade mark infringement (very costly!).

In addition, here are other reasons why you should not choose such a guerrilla installation. Because of the concept of single fault tolerance it is necessary in any domestic installation that all electricity is supplied at a point before the fuses and circuit breakers; if they are not, you might as well leave them out entirely. Since your fingers are not meant to touch cables with 230 Volts AC you will have to ask an electrician to do that job. The electrician will insist that there is a single cable from the inverter to the fuse box and he will insist that either a special plug-and-socket combination is used (so that you can't possibly touch any conductive parts) or that the inverter is connected to a fixed terminal.

The work of the technician will cost you something like 200 to 300 Euros which will extend the time for amortization by quite a few years (if you already did all the work and the electrician only has to insert the cables into the terminals it might be much cheaper).

I think it makes sense to explain why the electrician has to insist that the inverter has to have its own cable to the fuse box (in Germany and quite a lot of other countries it must; in all other countries it should). Let's assume we use an existing socket outlet on an existing cable for the connection; somewhere in the middle (between socket and fuse box) another cable branches to supply electricity to some other sockets. In the fuse box there is a 6 Amps fuse for the protection of this particular cable (it is an old house and they used thin cables at that time).

You switch on an appliance (or several) and they consume just a bit less than 6 Amps. Now you switch on the inverter; the sun is shining brightly and the inverter delivers 4 Amps. You have just found an old electric heater on the attic and you would like to know whether it still works. It's a model that consumes a maximum of 1 kW, so it will draw 4 Amps. You plug it in, switch it on and now we have a current of 10 Amps on a line meant to take a maximum of 6 Amps. The cable will get warmer and the question is how much warmer. Will it be enough to start a little camp fire within the wall? To prevent something like this to happen the electrician must insist that there is a cable from the inverter directly to the fuse box.

We come to the legal aspect again. The seller said that there is not a single law stating that you may not use installations like the one he sells. Strictly speaking he is right, but in the end he is utterly wrong. As already so often with photovoltaics, we need to have a closer look.

Indeed, there is not a single law stating anything about how electric installations must be carried out. However, when it comes to a court case there will be questions asked based on rules set by generally accepted technical standards. The standards are not laws as such but the judge will behave as if they were! Additionally there are the terms and conditions of the electricity company (just in case that the meter was indeed turning backwards then you may face a trial because of fraud and the tax office may well be after you because you were feeding into the mains grid and you did not pay any taxes for that).

Now we assume you had a little fire in the kitchen because you forgot to switch off one of the rings of your electric kitchen stove. The fat in the pan became really hot and caught fire. Before you were able to extinct the fire the flames consumed a cupboard as well. A good job that you signed a householder's insurance contract; you immediately phone the insurance company. They send an expert who looks at the kitchen. It is quite clear to him that fixed electric equipment caused the damage. Then by chance he sees your guerrilla solar installation. He takes a picture and bids you farewell.

The insurance company now claims that the electric installation did not comply to all relevant rules and regulations and therefore they would not be obliged to pay out anything at all (even the burned fat in the pan is of no interest any more). Ultimately you should be happy that you only had to pay for a complete renovation of your own kitchen and some furniture. If the block of houses had burned down you could well be bankrupt for the rest of your life (I don't write this in order to spread panic; the chances that a guerrilla installation causes any damage might be very low but the advantage it brings is pretty low as well).

This was again a bit over the top but I had to make my point. There is no law stating that you must not use such installations. You are absolutely free to do so. On the other hand you have to take the full responsibility for the installation. Feel absolutely free to decide whether a few Euros saved are worth that risk.

You can be absolutely sure that, if things go wrong, the seller is going to defend himself using teeth and claws; you were his cash cow and he doesn't like the idea that this relation might be reversed. So best will be not to buy such an installation in the first place.

To sum up: after reading this book you know far more than necessary in order to select the components for a solar installation. Most probably it will take you only a short time to find all the components you need at a much better price. If you don't feed the electricity produced into the mains (i.e. you build an island or peninsula installation) there is no risk of problems with the electricity company or the tax office; such problems are simply not worth it. If the connection on the 230 AC side of the inverter is done by a registered electrician then he will take the liability and not you. So it might be a very good idea that you talk with him before you buy the components; if you buy something he thinks to be dangerous he will simply not sign.

Just to tell how malicious some of these sellers are: a big German company (something like the German Radio Shack) sells these sets. Part of the package (but not connected to the cable) is a 230 Volts plug. Additionally there is a sheet stating which regulations to follow. This is not stabbing somebody, it is just holding the knife so that somebody will run into it! Pleasant!

20. Suggestion for a small installation

Preliminary note: I never tried out this installation myself but simply used the technical data published by the manufacturers. Strictly speaking it is just an academic exercise to show how an installation could be planed.

I was complaining all the time about the fact that there are no legal and cheap switches for automatically switching from mains electricity supply to the solar installation and back again. Well, in fact there are devices which have such a switch as part of their design. I'm talking about UPS (uninterrupted power supplies). With them we can do a trick borrowed from yachts.

Many yachts - especially the bigger luxury variety - have a 230 Volts AC on-board electrical system powered by batteries and/or generators. When the vessel moors in a harbor a cable is put and connected to the mains electrical supply on the quay. Since owners of yachts like to show off their technical toys to their visitors, this installation automatically switches to mains when a connection is made to the grid and switches to the on-board power supply the moment when mains is disconnected or unavailable. This switch is that rapid that you might perhaps see a short flicker of the lamps but all the other equipment (even the delicate electronics) is not effected in the slightest.

In fact they have installed a UPS on board. 'Normally' the electricity comes from the grid mains and in case it is disconnected (somebody unplugs the cable) the energy is taken from the batteries and maybe some time later the generators start.

The nice thing with a modern UPS is that there are quite a lot of settings one can make. One thing that can be changed is the priority; one can either select grid mains as the normal source of energy or the battery as the main source. As far as the battery is concerned one can select two voltage levels. The first is the voltage at which the UPS switches from battery to mains (if available) and the second is the cut-off voltage when AC production from the battery is stopped completely (in order to keep the battery healthy).

The main device in the proposed installation is an 'WT-Combi-S integrated power supply UPS' (there are rumors that the no load or low load power consumption of this device is relatively high; so perhaps you can look around for something similar). This UPS delivers up to 3 kW in the form of a pure sine wave which should be more than enough power for a normal household. Costing 769 Euros, such a device is not really cheap but I think that the price will go down significantly when many people order UPSs of this type (if the colleague from India is right one should be able to get them for 450 Euros or less).

We buy 12 solar panels Long Energy Modul 185 Wp at 111 Euro each for a total of 1,332 Euros. These panels have an open contact voltage of 44 Volts, they are monocrystaline and they have an efficiency of 14.4%. Each panel delivers in maximum 185 Watts so in total we have 2.2 kWp. The panels measure 160 x 80 = 1,3 m^2, so we have in total 15,6 m^2.

We will set up our installation in Düsseldorf again and PVWatts says that we can expect these irradiation data:

	Jan.	Feb.	Mar.	Apr.	May	June	July	Aug.	Sep.	Oct.	Nov.	Dec.
per m^2	0.82	1.99	2.49	4.16	4.46	3.63	4.22	4.36	3.08	2.43	0.92	0.80

The average over the year is 2.79 kWh per square meter per day when the panels are more or less perfectly oriented. With 14.4% efficiency we will get about 6.4 kWh per day from the panels. After deduction of the electrical losses we will have about 5 kWh per day.

A family of four will consume between 8 and 10 kWh per day; over the year as a whole, 5 kWh will not be enough. We have a look at the situation in May. The ratio between the irradiation in May and the average over the year is 1.6 so in May we can expect a harvest of 1.6 * 5 kWh = 8 kWh. In Germany in summer time one will use less electricity than in winter time so we might have full supply during the summer period.

Now we need to find out how big the batteries have to be. To make it simple we assume that we need the same amount of energy during the day as during the night (you will have to find out what exactly applies to your way of life). So in summer about 4 kWh will be fed into the battery (for use at night) and the other 4 kWh are consumed directly.

In a 100 Ah battery at 12 Volt you can store about 1.2 kWh. You will normally want to prevent unnecessary damage to the battery so we only use 0.8 kWh. For storing the 4 kWh we would need 5 batteries. Since we want to have a 24 Volt battery bank (the why you will see in a moment), we actually use 6 batteries. There is no real need to plan any time of autonomy because we have the mains on which to fall back.

The panels on the roof have a peak power of 2.2 kW. Each single panel has an output voltage of 40 Volts. If we tried to get all this energy down from the roof to the charge controller then we will have a current of 55 Amps which is rather a lot. Our first step is to connect two panels in line, forming a string at 80 Volts (we have to stay below 120 Volts for safety reasons so that this part of our installation is harmless if accidentally touched). Now we have a current of 27.5 Amps which is still too much (one can control this current but not with reasonably inexpensive devices).

Therefore we use the next trick which is that we divide the panels into three groups (each group of four panels is formed of two panels in line and two strings in parallel). Since it is 'legal' to connect as many charge controller in parallel as one wants to, we use three charge controllers for charging the battery bank. Each of the chargers is responsible for a maximum of 733 Watts at 80 Volts which produces a current of 9.2 Amps. Fine, that can be handled easily. Additionally we have the big advantage that we could install each of these three groups differently. Lately people have developed a preference for east-west oriented installations; the total amount of energy produced per day is lower but the distribution over the course of the day is more constant.

Now we come to the reason why we want to have a battery bank with 24 Volts. 733 Watts at 12 Volts cause a current of 61 Amps. That can't be handled with a normal charge controller (at least no cheap one). If we double the battery voltage we come to 24 Volts and a current of 30 Amps (I neglect the half Amp because we will never get 30.5 Amps in Germany in any case). For 24 Volts there are cheap charge controllers which can handle 30 Amps. Another problem solved.

We cross check whether the battery can cope with the currents. A 100 Ah battery can be charged with 10 Amps without getting too warm and without gassing. We have three charge controller and each of them can deliver 30 Amps; so we have 90 Amps at 24 Volts. We now have to differentiate between two situations. The normal situation is that half of this energy is consumed directly inside the house. The problem is that we can't be absolutely sure about that so perhaps we get the second situation, which is that all the energy has to be fed into the battery.

If half of the energy is consumed before it can reach the battery then there are 45 Amps left which are distribute evenly over the three strings. Then we have 15 Amps going through any battery and with that the battery can cope. However, if we don't have any electric consumption in the house we would have 30 Amps going into the batteries. That would not kill them directly but they would get much warmer then is good for them and they might be gassing strongly. Since we know that the actual cost of batteries is mostly determined by the energy throughput during its life time we decide to spend some more money (double the investment but half the wear-and-tear, so the only real additional expense is in the form of higher interest charges) and double the capacity of our battery bank. Now we have 12 batteries with a capacity of 12 * 1.2 kWh = 14.4 kWh.

Our last step is to check the life time of our battery bank. We take the good and reliable PzS batteries. They have a life time of 1,500 cycles of 80% DoD, the equivalent of a life time energy throughput of 1,800 kWh. Since we have 12 batteries the total throughput will be 21,600 kWh. We assume again that we have 5 kWh per day and that half of this energy has to go through the batteries which comes to 21,600 kWh / 2.5 kWh = 8,640 deep cycles or 23.6 years. We might decide to make the battery bank a bit smaller).

All the components were found in January 2014 in the Internet just as normal offers. So the prices are absolutely realistic.

12 qty. Long Energy Module 185 Wp á 111 Euro	1,332 €
Transport of Modules	100 €
3 qty. MPPT 30A controller 24V/780W á 150 Euro	450 €
WT-Combi-S integrated charger UPS	769 €
12 qty. 12V, 100Ah-Batterie, PzS á 100 Euro	1,200 €
Work and small parts	150 €
technical inspection AC-side	200 €
Total	4,201 €

Again we work with a loan running 20 years and with 5% interest rate.

Interest rate per day	28.7 cents
Back payment per day	57.5 cents
to pay per day	86.2 cents

The installation costs us 86.2 cents per day and delivers 5 kWh per day. Any kWh produced by our own installation costs **17.25 cents** which is about 10 cents cheaper than getting it from the electricity company. We save per year 182.5 Euro and during the planed 20 year life time 3,650 Euros (though we pay back the amount we borrowed). That means we can finance the next installation from the money we saved and don't need to borrow it the next time; then we get the kWh for **11.5 cents**.

I admit that I left out the efficiency of the batteries. When you know your numbers you can take that into account. Since the charging current will normally stay within the recommended limits the gassing will never be strong and the efficiency might be near to 90%. Only 2.5 kWh go through the battery and 10% of them are lost that is 0.25 kWh per day costing 4.3 cents. So no major changes to the numbers.

I would like to point out a specialty of this installation which makes sure that the batteries will have a long life. This little trick depends on using an UPS which works with a hysteresis. Our battery bank has a capacity of 14.4 kWh and we adjust the cut-off voltage in such a way that we still have about 8 kWh in the battery bank when swapping to mains. In case that there is a blackout we would have (with several restrictions) one or two days of autonomy.

The voltage when to swap from mains to battery is set so that the battery is always up to 90% of its capacity. During summer time the SOC will mostly be that high that the lower threshold will never be reached. During evening and night 4 kWh are taken from the battery and during the day the battery gets full again.

The interesting part is going to take place in winter time. Lets assume the battery is actually full. In winter (that is December and January) we can expect about 2 kWh per day from the panels. Since we need about 8 kWh in the household the battery has to deliver the missing 6 kWh.

Within less than 24 hours the lower threshold will be reached and the electric power is then coming from mains. It might take several days until the upper threshold is reached. The important point is that the battery was fully charged (no sulphate) before it gets emptied once again. In average the battery is all the time in a healthy state.

No matter whether summer or winter, the battery is fully charged at all times before it is used again. No chance for big sulphate crystals to form and the charging is relatively slow which in turn means that the efficiency of the battery will be relatively high and that there will be very little gassing which keeps the corrosion rate low.

If you wanted to have a UPS for your house, you would have to spend a bit more than 2,200 Euro for that and it would only cost money. When you spend another 2,000 Euros the installation will start to earn money! Stop, error, fault! No, it is not going to earn money! It is going to save money and that is quite a big difference. When you earn money (i.e. make a profit) you have to pay at least income taxes. When you save money (you simply do not spend it), then you don't have to pay taxes on that! It feels as a rise of your paycheck but it is no income. Great!

Addition: I meanwhile found in the terms and conditions of one electricity company a paragraph delimiting the time for UPS use to 15 hours per month. So check the terms and conditions of your electricity company before you start to build something like the proposed installation. If necessary, change the company!

21. Forums and help

In Internet you will find forums about nearly any possible topic. So there are forums about photovoltaics in - I guess - any given language. If you are relatively new to a topic then you will not know all the technical terms not to talk about all the abbreviations. Sorry, there is no known shortcut for this problem.

Sometimes one can find different forums about photovoltaics in the same language but with different intellectual levels. The problem then is: if you select the lower level you will (think you) understand most; however, you will find a lot of misleading and downright wrong information as well and you are (not yet) able to distinguish what is what. If the level is high the amount of correct information will be much higher but what use is it to you if you can't understand it? I would advice to stay with the forum with the higher level. In technology it is mostly better (and therefore much cheaper) if you don't understand something (and admit it) than to understand it wrongly (and insist that you are right).

If there are some forums to select from then you should browse through some interesting threads (that is the list of contributions concerning a subtopic). Prevails a rough tone? In many forums they give the number of contributions of the participants. If the tone is rough then these numbers will be relatively small because most people don't like it if they get insulted. So after a short time they will stay away. That you should do as well since those are no good places for newbies.

In many forums the participants react pretty sniffy when they get the impression that somebody just wants them to do the work having his feet on the table. Since you already made your way through this book you at least showed that you are willing to do the work though you possibly don't know how to do it; this makes a big difference. Working through the book did not make an expert out of you. So you will have doubts whether the installation you have in mind will make sense and will work properly. That's absolutely natural!

Now you will have to spend some more work and describe your (planed) installation. You start with where it is (city or area, orientation and inclination of the roof), which panels you want to use (manufacturer, size, efficiency, power output, short current, open circuit voltage), which solar controller you want to use and why (brand, type and technical data; possibly why you want to use exactly this one), which battery and which inverter (with the main technical data). And to all these items you should give the price you will have to pay (you might get good tips where to get that stuff much cheaper).

Now you could retort (and you would be right) that this is quite a lot of writing. My answer to this is that this work is even done by professionals (at least once if it is all the time the same kind of installation he sells or sets up). You don't have to do this work in one day; take your time and slowly you will be convinced that at least to you it all sounds reasonable. Now you are ready.

If you are not registered yet in that forum (somebody who is only reading normally does not need to) then you should do it now and start a new thread (don't use an existing one unless it has the title „my solar installation" or „what I want to build"). If you want to use components not commonly in use tell why (for example you get the panels for very little or no money). Then it is time to wait.

If in the forum they pay attention to good manners somebody (don't expect more than one) will send you welcome greetings. Then follow more or less objective critics (please keep in mind that the one who is less friendly or less objective can't possibly know you personally; therefore he can't insult you; if you answer in the same way an endless bickering will start; ignore the other one if possible and stay friendly even if it is hard to do so).

With the answers you get you have to distinguish which motivation might be behind it. Fantastic if you found someone who's only interest is to help and who really knows what he is talking about. You might get harsh critics but that is the best what could possibly happen to you! The next type are craftspeople and sales persons. If they write what you should do or have to do you can never know whether this is technically correct / necessary or whether they just want to sell you stuff (to find out the difference will be pretty difficult at times => stay polite and ask!).

Worst are the grousers and the ideologists. If you suspect someone to be of this type there is only one remedy: don't feed trolls! If someone writes that you are an idiot because panels of that type one must never connect with controllers of the other brand; ask politely why not. If you think the answer is technically not completely correct then you found a troll. Only way out is famish him.

If you can't find a forum that fits your needs then try to start a self-help group. The reason behind this is that one often understands things when trying to explain it to others (at university I overheard once a conversation of a student with his tutor; the student had written a pretty complicated computer program and had asked his tutor to listen to his explanations; after about ten minutes the tutor said that he was no longer able to follow these explanations but the student insisted to continue; half an hour later he said: now I'm sure that I did it right!).

So tell each other what you want to do, why you want to do it in this way and what you hope to gain. Certainty will come (or good ideas of how to check whether it is really a good idea). Often it is a good idea to start a pilot project with the whole group and everybody pays his contribution and participates in the planing and the work. If something goes wrong there is not a single person to blame.

22. Orientation of the panels

A few years ago this chapter would have been the most important one of the complete book. I put it now to the end of the main part of this book (further on are only supplementary chapters). The reason is the price of the panels. Three decades ago panels were produced one by one; two decades ago serial production started and one decade ago mass production started and the prices tumbled down rapidly. Now we are at the point that we can produce electric energy with photovoltaics cheaper than with any other technology actually known. And the prices are still going down year by year.

The prices of the panels don't play the dominant role they used to play. Some years ago, when setting up a photovoltaic installation, first the perfect orientation had to be determined; all the rest of the installation had to adopt to that (for example the under construction etc.). Nowadays you can't erect a cost-wise optimized installation without the extensive use of computer programs because you have to iterate through different values and see what the price-wise effect will be.

Or one uses the pragmatic approach: you don't look for the perfect installation but simply for a good one (as we saw with the irradiation data you can't anyhow know whether it is really the optimum you found; the basic numbers might be wrong to start with). Lets assume you want to set up an island installation. There are incredible programs by help of which you can compute how much energy you will harvest in the future (PVWatts V1 in case you live in the US or PVGIS if you happen to live within the EU). If you live somewhere else then the cards are stacked against you. Often you will not even be able to find out how much irradiation there will be at all during the months not to talk about a perfect orientation.

In supplementary chapter 4 you can find a short compilation of data from different sources. You can see differences of up to 20% or more between the different data bases (and they all claim to be right). I refuse to believe that the data of commercial programs are better than that.

Here is where the pragmatism comes into the game. Knowledge which is nearly right is much better than complete ignorance. So we use the data of the NASA because it is the only free data base with irradiation data available free of charge and for all places in the world. These data are in text format and it takes quite a while to understand how these data are organized (well, at least that happened to me); additionally you need a text editor that gives you the line number where the cursor is situated and you need a pocket calculator so that you can compute the line number where to look for the data you want (and the text file is pure ASCII and is about 3 MByte long). With the latitude and altitude of your place you go into the calculation, get a line number, navigate there and you got what you are looking for (?).

Well, I found that this way is pretty cumbersome; therefore I converted the text file into a file with binary data and I wrote a little program; you insert the coordinates and the program returns the data you need. This little program (it is called SolRad for solar radiation) is public domain and you can download it from my web-site www.jdhenning.de as a zip-file (program, code and data base are included). Then unzip it and use it.

SolRad runs directly on (as far as I know) all Windows installations; if not, you will need a compiler (and the rest of this story can't be part of this book). SolRad is a command line program (so it will run on the most simple computer you can think of). You can start SolRad directly (under Windows you just double-click on the name and a console will open) and the parameters are requested one by one (latitude, North or South, altitude and West or East). Or you open a console and start the program giving all the parameters in the command line.

What SolRad delivers is the so called horizontal irradiation. That is the irradiation that hits one square meter of ground per day. SolRad gives the average irradiation for all the months. Now we already know how much we might harvest if we put the panels flat on the ground (number of sq.m. * irradiation per day * efficiency of the panel = harvest per day).

We start with the easiest case: you live near to the equator. In the chapter about tracking systems we already had the fact that it is best to put the panels flat on the ground -that is 0° inclination. Out of practical considerations you will normally not wish to do that so they go onto the roof. In this area of the world the inclination of the roof is just high enough that rain water runs off well. How strong will the effect of this inclination be on our solar harvest?

The sine function gives an answer to this question. When we have a panel laying flat on the ground the sine gives us the irradiation falling on the area of the panel (1 means maximum; I could have stated 100% as well). At 0° there is the horizon in the East and at 180° is the horizon in the West. At 90° the sun is directly above us. So this little picture tells us the harvest over the day (with clear sky and no diffuse irradiation).

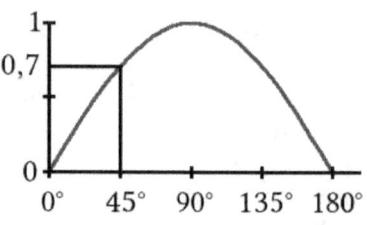

This sine function we can use the other way round as well. If we have a solar panel directly facing the sun (we have a 90° angle between the pointer to the sun and the panel itself), then the sinus tells us how much irradiation less we will have if we change the orientation to the left or to the right (that applies as well for up and down; then we would have to multiply the two sinus values).

So we are at the equator and our roof has an inclination of 20° (it does not matter in which direction) then we will still get 94% of the irradiation being possible in maximum (sine of 70°). Even if we had an inclination of the roof of 45° we would still get 70% of the irradiation. Now we have to include the fact that the sun (seen from Earth) is oscillating between 23° North and 23° South. Compared to northern Europe where East-West installations start to be popular we should have a 20° North-South inclination (while half of the roof brings full power the other side still generates more than 70% of what could be possible).

So near to the equator the alignment of the panels is no big problem.

In the chapter about the seasons of the year we had seen that the oscillation of the irradiation gets bigger and bigger the further we go away from the equator. Ultimately the reason is the ecliptic of the axis of the Earth (if that sounds interesting to you, you might want to read the last supplementary chapter).

PVWatts V1 uses as a standard for the inclination of the panels the latitude of the place. At the days when the night is as long as the day (equinox) then the sun is at midday directly above you in case you happen to be at the equator. If you happen to be in Hamburg (53° North;

10° East) that day, a solar panel facing south with an inclination of of 53° to the ground would be perfectly aligned to the sun.

Since the sun, as already said, oscillates between 23° South and 23° North, we aim at the longest day 23° too low and at the shortest day we aim 23° too high. So we get sine(90° - 23°) = 0,92 = sine(90° + 23°). In maximum we have a drop of 8% compared with the possible optimum for that day. As a pragmatist we can simply ignore this deviation (if that is not possible we just buy one panel more)!

If you want to get more energy during winter time then you have to increase the inclination (so you aim too low). If you want more energy in summer time you have to aim a bit higher. If you want to have the maximum harvest in Hamburg in winter time the inclination has to be 53° + 23° = 76°. Sinus(76°) = 0.97 so if you put the panel on the wall you only loose 3%.

There are great computer programs out there and they allow you to calculate any aspect of a solar installation but they come with a single disadvantage: they cost a few hundred Euros and it takes weeks to understand what exactly they are calculating (they normally don't tell you how and why they came to a certain result). Or you find a specialist for solar installations (and he has such a program and hopefully he knows what the different meanings are) and he will do these calculations for you for a few hundred Euros. So what to do if you can't effort that money or the time?

Think pragmatic! Most of the time you will have no choice where to put the panels anyhow because the house is already built. So you select the best part of the roof. An installation where the panels are flush with the roof will always be much cheaper than any kind of under construction especially made for you (the angle might not be perfect but with a perfect angle you will have to check that the panels don't cast shades at each other). The sine function gives you the difference in harvest and it is only a small calculation in order to know whether the additional effort will pay during the life time of the installation. Either there is a long and stony way with loads of calculations in front of you or you convince yourself that a good installation in a short time is better than a (possibly) perfect installation eventually in the distant future.

If that does not really calms you down then there is a trick you could use. Look in PVWatts V1 for a city being on the same latitude as the place you are living. In the first step you set the inclination of the panel to zero and let the program run. The result is the horizontal irradiation for one square meter per day for the other location. Then you use my little program SolRad and it tells you the data the NASA got for your place. Now you got the first correction factor.

Then you insert the local data (orientation of the roof, inclination) to PVWatts V1 and you let it run again for the other location. Now you can improve the result by help of the correction factor we generated in the last step (since the basic data might have deviations of 20% or more there is no way of getting reliably better data for your installation). Sorry, no simpler way for the poor man!

If you don't like calculations then you have the option to set up a small test implementation. For that you need a small panel, a solar controller, a small battery and a suitable consumer (car light bulbs which could consume a bit more energy than the installation can generate would be good) and you need an operation-hour counter. The battery can just be an old car battery (if it is not completely flat) and an operation-hour counter for 12 Volts you can get from Internet for less than 10 Euros. So this experiment is financially no big risk.

For our test set-up we make use of the fact that the solar controller will switch off the consuming devices whenever the battery voltage is too low and switches them on when the voltage is high enough again.

We connect the solar controller, the battery and the consuming device with each other and empty the battery until the controller switches off the consuming devices. Then we connect the operation-hour counter in parallel with the consumer, connect the solar panel and let the test start. Whenever the battery is full enough the consumer and the operation-hour counter are switched on and the time is measured.

After a few days or weeks we read the operation-hour counter. I just assume we had a 20 Watts light bulb in use. If we now want to know how much energy we harvested then we have to multiply the number of hours with 20 Watts. For example 20 Watts *100 hours = 2 kWh. Now we measure the size of the panel and search the documentation of the panel for the efficiency.

Now we got all the data we need in order to calculate how much power a real installation would have generated. This gives us a second correction factor (but to be honest: it is not of too much value since - at least in northern Europe - the deviations from year to year can easily be higher than 20%). The good message is that now you have a trustworthy bases for your planing.

Just in order to finish with this topic I have to have a word about diffuse irradiation. Depending on the latitude and the height above sea level the amount of irradiation which was at least once reflected on its way down to the surface of the Earth is different. Even the soil, the plants and the hills reflect some of the light (luckily, else it would be pitch dark in the house).

In Hamburg about 50% of the irradiation is diffuse. If you lay down a solar panel on the ground then this panel will get 50% of the global irradiation (as diffuse irradiation) plus the direct irradiation. So it is only a question of time until it makes economically sense to plaster even the north side of a house in Hamburg with panels (just a factor of four price-wise or even less; I personally think that this is only a question of ten years or so; yes, I'm an unstoppable optimist!).

1. Supplement: smart grid

The German 'Energiewende' did not just concern itself with looking at ways of producing as much electricity as possible using alternative sources but also at ways of saving electricity; a good example is the way minimum standards of energy consumption for different classes of appliances were set. A third approach was based on the idea of making a better match of demand and supply over time.

Till quite recently the ordinary householder used a very simple electricity meter, of the type in which a horizontal metal disc rotates as electricity is consumed. As far as he or she was concerned, running the washing machine at noon (when demand for electricity is at its highest) cost no more than running it at midnight, when demand is low. Now the new smart meter is being introduced, which uses the Internet (or other methods) to communicate with the computers owned by the electricity distributor and at the same time with all the appliances inside the household. So this smart meter knows that there are dirty clothes in the washing machine waiting to be cleaned; the electricity distributor tells the meter how much electricity costs at what time during the day and the meter then asks the washing machine whether it wants the energy at that particular price (this is a rather simplified explanation, but you get the general idea).

That such a meter will cost a bit more than the simple meter we used to have will be fairly obvious but it is assumed that we all are quite willing to do our bit to save the environment. It is also clear that a washing machine that can negotiate the price via the smart meter will cost a bit more than that old washing machine we used to have. The idea that we can flatten the daily peaks of electricity consumption by simply swapping the time when the electricity is actually being used has got something amazing.

Since everybody thought this was a brilliant idea (especially people in the electricity generating and distributing industry and of course most of the politicians) the EU decided that smart meters should be installed nationwide in all the countries of the EU, unless there was proof that the installations didn't make economic sense (EU-guideline 2009/72/EG electricity). Meanwhile, the electricity industry supplied studies showing how many billions of Euros could be saved in the economy of each nation.

The German Ministry of Economy and Technology asked Ernst & Young GmbH to make a study[1] which the ministry announces on its web-site:

> "The generation of electric energy from renewable sources makes it necessary to couple the grids, the production and the consumption in an intelligent and efficient way. Intelligent grids ("smart grids") shall balance the demand of electric energy with the fluctuating generation from renewable sources. Intelligent metering systems and counters could play an important role. The Ministry of Economy and Technology entrusted the auditing association Ernst & Young GmbH to furnish an opinion with an economic evaluation of a nationwide implementation of such systems and meters in Germany."

1) http://www.bmwi.de/DE/Mediathek/publikationen,did=586064.html

Stand by for a fanfare of trumpets! However, there are people who actually read such studies (me for example) and plow their way through 239 pages with economist jargon. Voilá, here is the result in short!

The effects on the electricity suppliers:

- Accounting at all hours and remote reading of the consumption

- Remote switch-off possible (somebody did not pay his bill)

- Conversion of running contracts to pre-paid contracts

- Lower costs for debt collection

- The costs for reading the meters goes down from 3 Euros per reading to 0.05 Euros

- Better forecast of consumption through the use of better statistical data (lowering the acquisition costs of electric energy by 6%)

- Shorter discussions with the clients (more transparency)

- Better grid-planing could save up to 2.5% of the total costs (of grid-planing).

Implications to the customers

Ernst & Young had a look at which parts of the electricity consumption could be moved to other times and how far they could be moved. Making coffee, switching on a reading light or looking at the news on a TV can only take place at the moment the customer wants to engage in that particular activity. That means that about 40% of private electricity consumption can not be moved to another time.

The other 60% equals about 85 TWh (Terawatthours) which is in fact rather a lot. Electricity consumption which can be moved at all (or at least some of it) is that used by fridges, freezers, washing machines, tumble dryers, dish washing machines and night storage heaters. They then weighted how many households actually have these appliances, how often they are used and the maximum number of hours the consumption could be moved.

	Electricity Consumption [in GWh/a]	Share of total Consumption (2011) [in %]	Movable Consumption [in GW]	Max. Share of Peak Consumption [in %]
Cooling	5,634	1.0	0.62	0.8
Freezing	2,112	0.4	0.60	0.8
Washing	1,175	0.2	0.58	0.7
Drying	1,964	0.4	0.63	0.8
Dish washer	4,041	0.7	0.60	0.8
Hot Water	2,681	0.5	0.68	0.8
Night Stor.	18,700	3.5	16.00	20.0
Heat Pumps	2,803	0.5	0.80	0.9
Sum	39,110	7.2	20.50	26.0

Ernst & Young found that 7.2% of the electricity could be used at another time if the households got better price information and used it as well. If we leave out the night storage heaters (see the next chapter), which had been banned some years previously but which were permitted once more, then only 3.7% of the consumption can actually be moved. The households use 25.5% of the electric energy in Germany (source Wikipedia) then about 15% of their consumption can be moved. Now an average family of four has a yearly consumption of 3,285 kWh (9 kWh per day), so that in fact a mere 493 kWh - worth all of 133.11 Euros - can be moved. Industry offered a 10% price cut for off-peak electricity, which means that the average family of four people can save **13.31 Euros** per year by using the smart grid.

The (German) Wikipedia article on smart meters states: "*According to the Deutsche Energie Agentur the price [for installation, in 2010] was, depending on the provider, between 35 and 100 Euros; and per year between 60 Euros and 240 Euros.*" [The Deutsche Energie Agentur (German Energy Agency) is a company owned by the state, a number of banks and insurance companies]. Ernst & Young estimated installation costs of 75 Euros and an annual cost of at least 120 Euros. The new meters do not last as long as the old ones did (a mere 12 years before they need recalibration); it seems reasonable to assume that using smart meters and a smart grid will cost **130 Euros per year per household**.

In addition, the meter itself uses energy (the following calculations assume a cost of 27 Cents per kWh). The old type of counter (the one with the revolving metal disc) consumed 3.4 Watts when any electricity at all was being used in the house, which comes to a maximum of €7.09 per year. A smart meter uses 15 Watts at all times, so the consumer has to pay €35.47 per year. In addition, although I did an intense investigation [I even read the the data sheets of some meters] I could not find out whether the meter includes its own consumption), which leads me to conclude that the client will have to pay for this consumption. So just powering the meter costs the customer an extra **27.44 Euros per year**.

Upshot

To put it simply, the client is expected to **pay 160 Euros** per year in order to **save 13.31 Euros** - and that with a countrywide coverage of 80% by 2022. Somehow I was reminded of the time when commercial TV was introduced in Germany. The RTL channel became well-known for its remarkable low intellectual level or, as many people said, "It is not possible that they can sink even lower!". But they could, and they did, again and again, over the years. Every now and then I used to think that this or that politician could not possibly be quite so stupid, but I am faced by the fact that they can and they do again and again!

That is not all; they are not going to stop at this point. It is expected that in the near future there will be more and more electric cars on German roads and they want to include all these cars into the concept of the smart grid. I'm going to leave out all the costs of the infrastructure (I estimate that the necessary investment will be gigantic and that it is the ordinary householder who will have to pay for it all) and only have a look at the energy storage. In the chapter about batteries we saw (or will see) that to store one kWh in a battery and to get it out again costs about 10 cents (the manufacturers of the special batteries used in electric cars expect to reach this price in - hopefully - 10 years time; at present it is between 50% and 100% higher than the ordinary lead acid battery).

By offering a discount of 10% on my electricity bills, the electricity generating industry would actually pay me 3 Cents per kWh for using my battery in a way that costs me 10 Cents per kWh. I have no idea how they will introduce such an expropriation in the German Constitution but I'm sure that there will be quite a hard struggle.

Or we could just change our point of view. If you wanted to shoot down the whole concept of the 'Energiewende' what better tool could you find than the smart grid? It makes all the energy (except the traditional way with coal and nuclear power plants) extremely expensive and people will start to hate solar panels and wind turbines. One simply can't think of a better way.

I would like to leave the last word on this topic to Ernst & Young:

> **The EU concept is neither economic nor practical for Germany**
>
> *The EU concept of a compulsory change to the new intelligent metering system with at least 80% coverage in Germany by 2022 will:*
>
> - *have no economic advantage because of the negative net capital value*

- *force the greater part of households into paying excessive costs*

- *create considerable practical problems in setting up a largely untested system at too fast a pace*

- *require very large investment with a high and unknown risk factor*

- *not lead to a long-term sustainable change with benefiting factors since EEG installations are not included whereas small consumers (households) will have to contribute far more than they can save.*

I tried and tried and tried once more, but I simply could not stop myself - I have to have the last word. The politicians probably know that they are making a mess but they continue to do it anyhow! There were times in the past when it was not possible to distinguish whether the German nobility was controlling Spain or whether the Spanish nobility was controlling Germany. Somehow I begin to have really bad suspicions!

Amendment: I was told as a child that it is not a sign of good education if somebody tries to have the last word all the time. Luckily I found some nice information on the web-site[1] of the CKW (the power plants for the central part of Switzerland) dated from 13 January 2014. They did a 3½ year project in order to study the effects of using smart meters. About 1,000 households took part in this study; 400 households volunteered and were very interested in saving electricity and 600 households were 'pressed into service'.

1) http://www.ckw.ch/internet/ckw/de/medien/news/archiv/2014/smartmetering-pilotprojekt.html

The central two sentences in this publication are: "The benefit of smart meters is coupled [strongly] to the determination of the client to change his long term patterns of consumption. If this [determination] exists then smart meters could make sense at times." They do not even use the word 'can', instead they use the word 'could'. Everything clear? Oh shit, I did it again. Sorry folks.

2. Supplement: night storage heaters

This way of warming your home had stopped being used in Germany, when in 2013 the German Government decided to permit this type of heater to be used once more. This caused quite a lot of protest from the environmentalists.

In order to understand why electricity companies think that night storage heaters are essential and why environmentalists think of them as the work of the devil, a little explanations seems necessary.

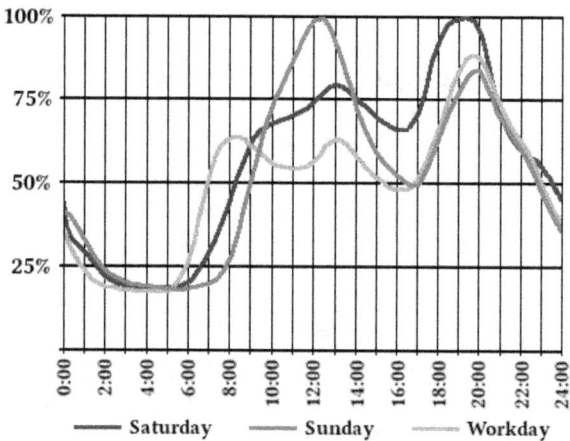

We begin by having a look on how much electricity is consumed in Germany during the day. We see that there is a large valley between 10 p.m. and 7 a.m (shown on the chart as 22:00 and 07:00). At noon and during the evening there are high peaks - the exact time depends on the day of the week (source: own work based on Wikipedia; public domain).

Very broadly speaking there are three types of power stations producing electricity for the grid. There are base load power plants which run day by day with a constant output (low personnel costs and high profits), power plants for mid-load which can flatten the big bumps in the curve and lastly power plants for peak loads which run just a few hours per day (they have high personnel costs and also high operational costs). This situation seems to make the idea of night storage heating a very sensible idea. However, just to put into perspective what is involved: in

Germany a total of about 600 TWh of electricity is produced each year and night storage heating consumes 10 TWh or 1.6%.

These heaters are in principle just a large metal box filled with fire-clay blocks. During the night the blocks take up heat and in addition heat the home (with the help of a small ventilator). During the day the blocks retain enough heat to warm the house till the evening. In order that the customer will think that this kind of heating is economical, the electricity used for running these storage heaters is a few cents cheaper at night. This made the electricity companies a little happier because that big dip during the night was partly evened out, at least in the winter time (why storage heaters help in any way during summer nobody has managed to explain as yet).

What gets the environmentalists upset is that power plants using fossil fuel have an efficiency of just a little over 30%, whereas ordinary domestic gas central heating systems can have an efficiency of almost 90%. Because of this gigantic wastage of primary energy, night storage heaters were banished (though with generous deadlines and many exceptions).

On the 17th of May 2013 this ban was repealed. This happened because somebody had the bright idea of storing the surplus energy of wind turbines and solar installations in storage heaters. It would no longer be necessary to limit the output of these power sources in case of overproduction which would make a noticeable difference to the national economy.

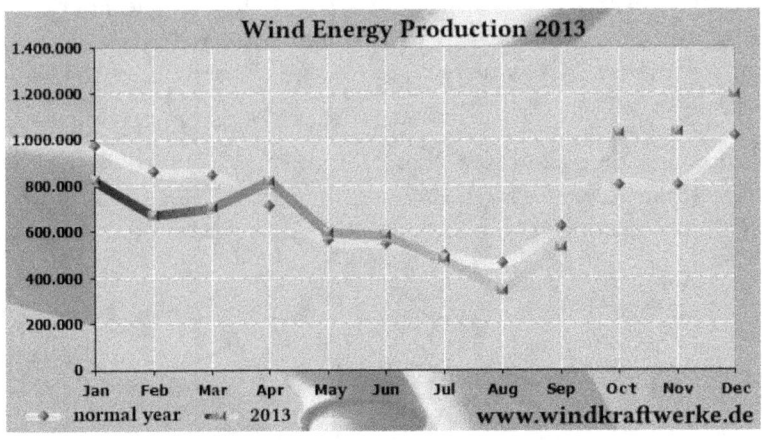

This diagram shows the amount of electricity produced by wind turbines over the year (Germany inland). Between May and September only 50 to 60% of the maximum possible output is actually produced. Then output increases till December after which it goes down again. It would seem that heat storage units could indeed take up the surplus electricity generated by the wind turbines.

However, if you take a good look at where most of the big wind turbines have been built (that's anywhere from near the coast to many kilometers out over the sea) on the one hand, and on the other hand where the homes are that actually need heating (i.e. densely populated areas), then some critical thoughts might arise.

Purely by chance, I stumbled across some information published by the "Ministry of Energiewende, Agriculture, Environment and Rural Areas of Schleswig-Holstein" (Schleswig-Holstein is the most northern part of Germany). The publication stated that in 2011, looking at all of Germany, some 420.6 GWh of electric energy were lost because wind turbines were switched off, and 80% of these losses were directly affecting Schleswig-Holstein (source of information: National Grid Agency; Monitoring Report 2012). And the reason for turning down the wind turbines was . . . . that the energy could not be passed on to the next high tension grid!

That would mean that about 0.07% of total electricity production was simply lost because it could not be passed along. In 2012 the total amount of electricity produced by wind turbines came to 50.7 TWh. So those 420,6 GWh are just 0,83% of all the power produced by wind turbines. If you know that about 6 to 7% of the electric energy is lost during transmission along the cables (conversion to heat), then this glorious saving of all this lovely wind energy is not being very realistic.

And if you have a look at the next chart (http://www.volker-quaschning.de) then you see that not only do photovoltaics have a pronounced variation through the day; the same applies to wind energy. The maximum produced can be as much as four times the minimum.

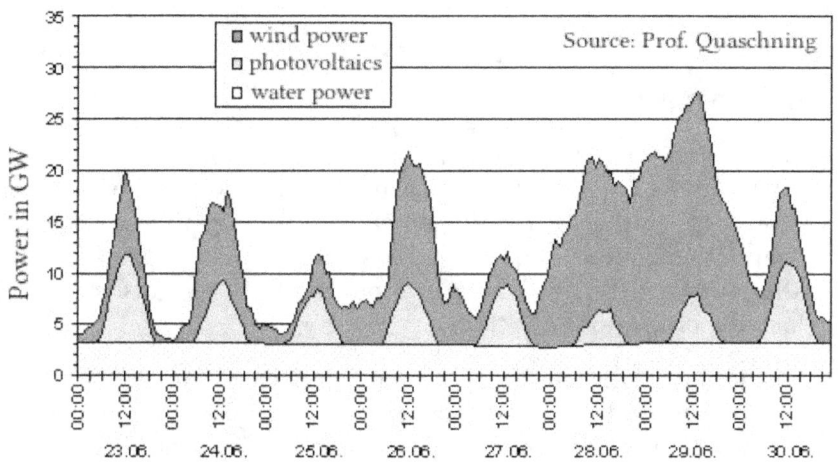

Anybody proposing to use the electricity produced by wind turbines or solar panels for heating up (night) storage heaters has definitely no clue as to energy production and distribution or he is possibly following different aims (maybe both or did you ever see storage heaters on a dike?)!

3. Supplement: measuring the consumption of appliances

From the chapter about physics we know that power can be calculated using the formula $P = V * I$. So if we want to know how much power our fridge needs when the compressor is actually running, then we open the circuit at a suitable location, we use a screw terminal to make contact with the probes of the multimeter, we put the fire extinguisher in place and push the plug into the socket.

Such a solution might work for a trained electrician but you should not try it; you must just forget the proposal instantly and completely. So how can you find out how high the consumption of your fridge is?

There are simple (and safe) devices for measuring the consumption. These look like one of those plug-in time switch units. You plug this device into a wall socket and then connect the cable of your fridge to the socket of the measuring device. Without a display such a measuring device would be pretty useless and most of them have a LCD (liquid crystal display) panel. There are so many makes out there that it becomes impossible to give a complete description so I hope for you that the manual that comes with the meter is understandable (if not, I can at least give you some idea what you most probably will find). However, firstly I have to make it clear that such a device can be used by everybody without any danger whatsoever!

Such measuring devices are sold every now and then by discounters (at least in Spain and Germany); else you have to look for them on the Internet. The prices start at about 9 Euros and go up to (at least) 40 Euros. Most will cost 12 to 15 Euros and they are the ones we want; it is almost certain that they can do what is necessary.

At this point I would like to make a remark to the manufacturers of these devices. If I want to measure the AC consumption of an appliance then there has to be tension on the line, so a power supply for the measuring device is guaranteed. However all these measuring devices I have come across have two little batteries built in which cost two Euros each if you have to replace them (after a year). It could be that it is the manufacturers of these batteries who also produce these measuring device. Hey you guys, why don't you produce decent devices which have a little power supply built in and dump the batteries? That does not even cost more!

When you switch on the measuring device (if it is being hesitant in showing figures, try taking out the batteries and inserting them again) a measurement cycle is started. It is possible to display the actual voltage on the line, the actual current flowing at the moment and how many Watts are being actually used. In addition they show on the display how many hours the measurement cycle has been running and how much was consumed during this period (in total and often as well per hour).

Some measuring device can show additional parameters or they even have a connection to Internet (but then they cost a lot more than 15 Euros) so that you can get all the values from Palma de Majorca while you are having a holiday in Hawaii, although I feel absolutely sure that you don't need the data there. In the end we are only interested in finding out how many Watts the fridge of Aunt Gertrude is taking when the compressor is on or how much the old PC needs (and that we don't need to know by the minute with table and diagram).

Most of these devices are exact enough for even measuring the standby consumption of computers and TVs. If you are thinking about swapping to solar power only these measurements are of importance to you. Whereas it will be very difficult to find information in the Internet how much your deep fryer consumes when you serve yourself a plate with french fries. With such a meter you will know that at the moment you start to eat.

4. Supplement: comparing NASA, PVWatts and PVGIS

It is a pity that there are hardly any databases with worldwide climatic data which we can open without paying for the privilege. In my investigations I have found just one single database with such data that is free. It is a data base published by NASA; you just introduce the coordinates of a place and you get the irradiation data for every month. A database in which you can get similar data for selected places in the US is PVWatts; for the rest of the world its data are scarce. The PVGIS database only contains data for Europe and was published directly by the EU, so the data should be pretty reliable.

In the following tables you see values for so-called horizontal irradiation; this is the radiation falling on one square meter of horizontal ground per day. The purpose of these tables is to show that even official data differ quite a lot; for the places out of Europe I calculated the percentile deviation.

```
      Jan   Feb   Mar   Apr   May   Jun   Jul   Aug   Sep   Oct   Nov   Dec   Avr

Stockholm 60n 18e
      0.33  1.00  2.38  4.20  5.86  6.22  5.90  4.55  2.88  1.29  0.49  0.19  2.94   (NASA)
      0.26  0.77  1.78  3.75  5.29  5.36  5.06  3.81  2.32  1.17  0.44  0.20  2.53   (PV Watts)
      0.29  0.90  1.98  3.59  5.30  5.38  5.22  3.88  2.47  1.16  0.44  0.17  2.58   (PVGIS)

Hamburg 53n 10e
      0.68  1.32  2.34  3.62  4.74  4.77  4.66  4.06  2.73  1.55  0.80  0.53  2.65   (NASA)
      0.49  1.12  2.25  3.80  4.96  4.93  4.60  3.98  2.62  1.51  0.72  0.36  2.62   (PV Watts)
      0.56  1.20  2.50  4.28  5.20  5.42  5.11  4.24  3.01  1.69  0.71  0.52  2.88   (PVGIS)

Munich 48n 11e
      1.17  1.97  2.97  4.12  5.08  5.27  5.27  4.68  3.26  1.96  1.15  0.88  3.15   (NASA)
      0.88  1.66  2.53  4.03  5.51  5.02  5.61  4.70  3.13  2.08  1.03  0.67  3.08   (PV Watts)
      0.98  1.73  2.90  4.48  5.03  5.38  5.24  4.47  3.30  2.05  1.15  0.82  3.14   (PVGIS)

Paris 49n 2e
      0.95  1.62  2.64  3.87  4.93  5.19  5.21  4.58  3.18  1.91  1.11  0.73  2.99   (NASA)
      0.78  1.39  2.29  3.63  4.61  5.31  5.35  4.85  3.11  2.02  1.04  0.61  2.93   (PV Watts)
      0.95  1.61  2.95  4.62  5.30  5.83  5.55  4.75  3.64  2.09  1.12  0.78  3.27   (PVGIS)

London 51n 0e
      0.82  1.46  2.45  3.72  4.71  4.97  4.98  4.34  2.93  1.79  0.99  0.62  2.81   (NASA)
      0.71  1.20  2.12  3.64  4.92  4.91  5.01  4.35  2.97  1.74  0.97  0.55  2.77   (PV Watts)
      0.77  1.40  2.59  4.17  5.08  5.51  5.22  4.27  3.20  1.87  0.99  0.62  2.98   (PVGIS)
```

218

```
           Jan  Feb  Mar  Apr  May  Jun  Jul  Aug  Sep  Oct  Nov  Dec  Avr

Madrid 40n 4w
          2.14 3.06 4.55 5.69 6.63 7.52 7.55 6.64 5.03 3.49 2.31 1.87 4.71 (NASAfigures)
          1.97 2.78 4.44 5.37 6.18 7.14 7.42 6.37 4.47 3.27 2.22 1.44 4.43 (PV Watts)
          2.07 3.15 4.49 5.69 6.60 7.74 7.98 6.99 5.37 3.59 2.37 1.92 4.84 (PVGIS)

Bombay 19n 73e
          4.97 5.66 6.40 6.72 6.56 4.62 3.57 3.51 4.35 5.18 4.93 4.58 5.09 (NASA)
          4.53 5.32 6.20 6.86 6.56 4.84 3.77 3.84 4.20 5.10 4.73 4.27 5.01 (PV Watts)
             9    6    3   -2    0   -6   -5   -9    4    2    4    7    2 % difference

Rivas (Nicaragua) 11n 86w
          5.45 6.02 6.78 6.73 5.83 5.40 5.32 5.41 5.17 5.13 5.04 5.13 5.62 (NASA)
          5.11 5.81 6.10 6.04 5.33 4.52 4.74 4.55 4.49 4.30 4.33 4.54 4.98 (PV Watts)
             6    4   11   11    9   19   12   19   15   19   16   12   13 % difference

Mombasa 4s 40e
          6.24 6.57 6.51 5.60 4.80 4.75 4.91 5.63 6.30 6.40 6.27 6.15 5.84 (NASA)
          5.78 6.05 6.03 5.31 4.51 4.66 4.62 5.14 5.71 5.88 5.83 5.50 5.41 (PV Watts)
             8    8    8    5    6    2    6    9   10    9    7   12    8 % difference
```

5. Supplement: using a multimeter

As with all skilled crafts you need special tools and the same applies to photovoltaics. The most important one is the multimeter and we only need the very simple one which you might find in any junk shop (Pound Shop, Todo Cien, Chinese Bazar or whatever they are called). Or you could have a look on the web. There are many different manufacturers and therefore my explanations might not be absolutely correct for the device you will buy, but the main functions will be there.

Such a multimeter is a bit bigger than a pack of cigarettes and the simple ones cost about 10 Euros. In the upper part we have a display; below this display is a big rotary switch in the middle of the multimeter, with a lot of symbols and stuff at the edge which might not mean very much to you right now. Under the rotary switch there will be two or three sockets; these are for the cables with the probes. Finally, there are two cables, a black one and a red one. Over the last few decades it has become common practice to use the black cable for 'minus' and the red cable for 'plus' (and you should follow this convention because it saves you a whole step when thinking about a technical problem which is already difficult enough).

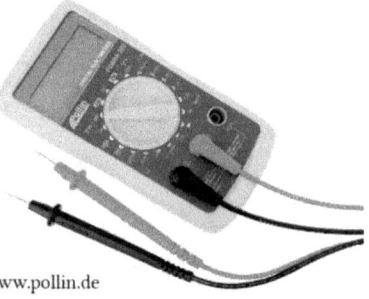
www.pollin.de

We connect the black cable to the socket labeled COM, GND or with a small symbol like the one in the little picture. The red cable we put into the socket labeled 'VΩmA' or something similar.

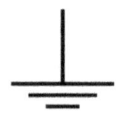

As long as the test prods are just resting on the table you can turn the big rotary switch back and forth as much as you like, no harm will be done. As long as both prods are in contact with a circuitry don't turn the switch (first disconnect at least one prod). When you work with a multimeter you will again and again come into situations where you simply haven't got enough hands - with one hand you hold one probe, with the other hand you hold the other probe, with the third hand you hold the multimeter and with hand number four you switch on the device.

That's why most multimeters have a very helpful function, namely the continuity tester. With its help you can detect whether any two points in a given device have a direct electric connection. When you use this function, the rule is: switch the device off AND pull out the plug! Now you look at the symbols around the rotary knob to see whether you can find a little symbol consisting of a point with some concentric circle segments. Turn the rotary knob to this position and put the tips of the probes together and you will here a beep. No beep, no electric connection. This function substitutes for at least one hand and additionally you don't have to look at the display (a friend of mine delved deep into his junk-box and unearthed two multimeters which behaved differently; with one the symbol was more like an arrow pointing to a plus-sign (that is the symbol for a diode); and the other produced no beep but displayed a 0.01 when there was a contact [not too practical]).

Now we make the first experiment that makes real sense: we measure the voltage of a battery. You turn the rotary knob to the DC-V field ('DC' for direct current and 'V' for voltage) and then on to the 20. Now you contact the minus pole with the black probe and the plus pole with the red probe and now you can read the voltage of the battery. Next, you swap the probes. You will read the same voltage as before but now there will be a minus sign in front of the number. That is, we can not only find the voltage but also the orientation of the electric source.

DC voltages higher than 120 Volts and AC voltages higher than 50 Volts are not just potentially dangerous, they really can be deadly. Use the other settings of the rotary knob only when you are absolutely sure that you know what you are doing. Don't forget, every time you have finished using your multimeter, switch it off or its battery will be flat when you most need the meter.

Tip: there are crocodile clips available which can be directly connected to the probes of the multimeter. If you use them you might have two more free hands.

6. Supplement: solar cooling

This chapter is addressed mostly to the industry (except the part about modifying existing fridges) since fridges and freezers especially designed for solar power can not be found on the market (at least not in numbers).

Naturally it is understandable that a company prefers to produce something after the competition proofed that there is a big enough market. It is not that I only think that there is a huge market, I can proof it. It is a study about the energy revolution in Cuba[1].

1) http://www.oe2.de/fileadmin/user_upload/download/Energierevolution_Cuba_dt.pdf

Cuba has about 11.3 million inhabitants and during this 'revolution' 2.5 million old fridges were replaced by modern and effective fridges. The state did not do this out of friendliness; there were economic reasons since Cuba is not a rich country and they have to import petrol. Now one can make a projection: if already in Cuba there was one old (and power hungry) fridge for four persons how big is the world market for solar driven fridges which cost much less to run?

Fridge

In the chapter on fridges in the main part of this book we saw that it takes quite a lot of effort to keep a fridge running at night as well as during the day. So much effort was needed that we reached a price equivalent of about 30 cents per kWh (price to run the fridge two days). How then does this fit in with the statement made at the beginning of this book that electricity prices of 2.6 cents per kWh are possible? Very simple: 2.6 cents **if** the sun is shining and the power is used **immediately**!

Now a fridge needs to run 24 hours per day. Every kWh we pass through the battery costs us ten cents for wear and tear. On top of this come the cost of a safety margin (the solar panel is twice as big as is absolutely necessary and the solar controller is also bigger). This is the price you have to pay if you want to (or have to) adapt appliances, designed to run on a traditional electric supply, to run on solar installations instead.

However, we can do it all in a different way. If you dig into the history of fridges you will find out that up to 1950 it was not unusual to cut ice blocks in winter time and use them during summer in ice boxes, into which these blocks fitted. This principle is very old; already 2,000 years ago compacted snow was shipped from Spain over the Mediterranean Sea so that very rich people in Rome could enjoy ice cream.

A really modern solar fridge simply needs to be constructed differently. As a medium for storing the cold we use water (we want a fridge, not a freezer, although that would be possible as well). Water is the perfect medium for storing heat or cold and in addition, it is very cheap. Part of the space inside the fridge must be sacrificed to accommodate an insulated water tank so that the solar fridge will be somewhat bigger than a normal fridge. When the sun shines, the content of this tank get cooled down to slightly above 0°C.

I just make a sketch to show how such a solar fridge might be designed. First we need to have a solar panel. It needs to feed 50 Watts to the system in order to meet the requirements of the fridge itself. Additionally we need to cool the content of the tank, which is our reserve for times with little irradiation. So we select a 75 Watt solar panel. It would be stupid not to use the heat pump effect of a compressor driven unit, but this time we take a DC motor which can start reliably without high starting currents. In order to keep the losses low we use an MPP solar controller.

A normal standard fridge is about 60 by 60 centimeters on plan (0.36 square meters) and is nearly 90 cm high. We take the insulated water tank as a plinth for the fridge; the water tank will be approximately 50 centimeters high. Such a tank would have a capacity of about 150 liters.

Water has a specific heat of $4.182 \text{ J} / (\text{g} * °F)$ which can be calculated to be $1.16 \text{ kWh} / (\text{m}^3 * °C)$; that means that it takes roughly one kWh in order to change the temperature of one cubic meter of the water by one degree. If we want the temperature in the fridge to be less than 8°C, then the water for cooling in the tank should have a temperature between 0°C and 7°C (we need one degree difference as a minimum for heat transfer to work). Now we can store about one kWh worth of cold in the tank (150 liter * 7 degrees = 1,050 Watts). It all checks!

Ho! Ho! Beginners fault! We made a big mistake. One kWh stored in a battery is not equivalent to one kWh of stored heat or cold in a tank. We left out the heat pump effect. The heat pump in a fridge converts the one kWh of electric energy into three kWh of thermal energy (the reason why air conditioning has a better coefficient of power is due to fact that the difference of temperatures is smaller there). We need a redesign!

The tank as a plinth for the fridge is not big enough so we need to double the capacity of the tank we first thought of. Now we have a tank as big as a normal fridge with a capacity of 300 liters and we can simply place it next to the fridge (it could be designed to look the same, which is fine). However, we are still only half-way to our goal. This problem we can solve with the use of alcohol. If you pour methylated spirit (alcohol that has an extremely bitter taste added to it, even an alcoholic wouldn't like it and which costs about 50 cents per liter if bought in quantity) into the water, the freezing point of the liquid drops below zero. If we replace 20% of the water by spirit the freezing point will be at -10°C. Now we have a potential temperature difference of 17°C and that is enough.

Water in the quantity required costs next to nothing. A well isolated water tank of the necessary size might cost 50 Euros (the price of a normal fridge if we leave out the costly intestines). To make it possible for the cold coming from the compressor to cool the water we need an additional heat exchanger; something like those you find at the back of a fridge but smaller. That I calculate generously to be 10 Euros.

Now we need to get the cold from the tank into the fridge. For that we need a little pump (about two Watts) to drive the water through the heat exchanger which has to be in the fridge itself in any case. This pump has to work during the night as well, so we need a small battery including some electronics; pump, electronics plus battery I estimate at 25 Euros.

Now we have the additional costs for a solar fridge:

Solar panel 75 Watt	75 €
MPP controller	45 €
Water tank	50 €
60 liters of methylated spirit	30 €
Heat exchanger tank	10 €
Pump & accessory	25 €
higher costs DC compressor	30 €
additional costs	265 €

Again we calculate with a credit over 20 years and 5% interest rate.

Interest rate per day	1.8 cents
back pay per day	3.6 cents
Daily payment	5.4 cents

This price we have to compare with the price we would normally have to pay for mains electricity. Since the electric consumption of a normal fridge comes to 0.5 kWh per day, we have an equivalent of **10.8 Cent per kWh**.

At this price a solar fridge would be competitive in nearly the whole world even if placed directly next to a mains socket. In some countries this would not be the case, as for example in many parts of US. The Americans will just have to wait until the panels and the other components for solar installation get even cheaper or the price for electricity from mains goes up a bit. But the gap is not really big any more.

Air conditioning

If the trick works so well for fridges, why not use if for air conditioners? The big advantage of the air conditioner is that we don't need much cooling power when the weather is bad, so that we don't need to bridge long periods of time (just a few clouds so now and then, or perhaps a maximum of a complete night). We will first try a system without alcohol.

Room climates with really low temperatures are dangerous for your health so that I assume that the air coming out of the outlet to the room should not be far below 20°C. So we can work with a temperature range of between 2 and 18 degrees for storing the cold without the risk of getting frozen water. Since we only have to bridge a single night this looks promising.

We design the following system: we have some solar panels connected to an MPP type controller and the DC-motor of the compressor is directly connected to the controller. The motor drives the compressor of a heat pump directly and we put the cold into a thermal storage (i.e. a tank with water). By help of a second circuit we move the cold from the storage to the heat exchanger of the indoor unit of the air conditioner. There air streams over the heat exchanger and cools down the air of the room.

An air conditioning unit with 2.5 kW thermal power needs 400 Watts of electricity, so that we need 9.6 kWh of electricity over 24 hours. This amount of electricity we need to capture and store during the day and convert directly into cold which we can store. Since the MPP controller has an efficiency of about 95% we will need 10 kWh from the panels. Again we take the data of Majorca. In July we can expect 6.72 kWh per day per square meter. The panel again will have an efficiency of 15%; so one square meter will deliver roughly 1 kWh per day. We need 10 kWh per day so we have to install 10 m^2 of solar panels.

Again we take the small 0.6 m^2 panels for 86 Euros; we need 17 of them, which will cost 1,462 Euros. A suitable solar controller we might get for about 100 Euros. A normal 230 Volts 2.5 kW thermal power air conditioning unit costs 350 Euros. The additional expenses for having a DC motor instead of an AC motor will be less than 150 Euros. Now we have to find out how big the cold storage needs to be.

During the day we get the cold into the storage. We have a CoP of 6 so that we generate 60 kWh of cold every day. What is needed during the 8 hours of the day is used directly. The amount of cold to be stored for bridging the night is reduced to 40 kWh. We remember that, roughly speaking, it takes 1 kWh in order to heat (or cool) one cubic meter of water by one degree. We have a temperature range of 18°C to 2°C = 16°C. That means that we need a volume of 2.5 cubic meters of water to store the cold for the night. That could be done but is a bit elaborate.

If we lower our expectations a bit and only request cooling power during the first four hours of the evening then the tank could shrink down to 600 liters. Well, that is still too much. We will have to use the trick with methyl alcohol once again. We replace 20% of the water with alcohol and we get a temperature range of 18°C to -10°C = 28°C. If we now want to have cooling for the complete night we would need an isolated tank holding 700 liters; that would be a cube measuring 90 cm in each direction and could be done without major problems. The version with cooling during the evening only needs 175 liters; that would be a cube of

90 x 40 x 60 centimeters, the size of a shoe cabinet. One could live with that.

Now let us have a look at the financial side of it all.

17 solar panels at 86 Euro	1,462 €
Solar controller	100 €
Add. expenses compressor	150 €
Water tank	50 €
60 liter meth. Spirits	30 €
Heat exchanger tank	10 €
Small pump & parts	25 €
Total	1,827 €

Again we calculate with a loan over 20 years and 5% interest rate.

Interest rate per day	12.5 cents
Back payment per day	25.0 cents
Daily payment	37.5 cents

A regular air conditioner, running 12 hours a day, consumes 4.8 kWh; since we have taken our example within Spain each kWh costs 22 cents, so we would have to pay 1.06 Euros per day. Our installation with the thermal storage runs for an equivalent of 7.8 cents per kWh, so we would pay roughly one third. By way of added compensation we have an aesthetically pleasing cube (for free!) perfectly suitable for displaying art objects.

Note: I did not account for the fact that the CoP of the heat pump will go worse when the temperature in the cold storage is lowered.

Cooling: can we do better?

One effect is known in physics which could be used and which you have already come across in this book. In the Desertec project they considered using phase shift storages for storing heat so that it could be used during the night for the production of electricity. We are not going to melt salt; we are going to melt ice, and we need salt so to set a suitable melting point. First, let's have a look at where that might get us.

Ice has a melting enthalpy of 333 kJ/kg which corresponds to 92.5 kWh/m^3 (assuming that we neglect the different densities of water and ice). That means that if I have a tank with one cubic meter of ice at 0°C I will need to put 92.5 kWh in the form of heat into the tank before the temperature starts to be higher than 0°C.

In our last version of a solar fridge, we had to use a tank containing 300 liters of a water and alcohol mix, and we worked with a temperature difference of 17 degrees. In such a tank we were able to store 5.1 kWh of cold. Well, now we are going to change the design once more. The new design means that we fill simple containers (plastic bottles) with normal water from the tap and put them into the tank before we fill it with the water-alcohol-mix. In this way we replace about 200 of the 300 liters with normal water (we can't replace all the liquid because it is needed for transporting the heat/cold).

And now comes a little surprise! In addition to the 5.1 kWh we already were able to store, we can now store another 18.5 kWh of melting heat/cold, so that in total we are able to store 23.6 kWh of cold. With a simple fridge we consumed 0.5 kWh of electricity, which in fact equaled 1.5 kWh of thermal energy because of the heat pump effect (COP of 3). We had calculated that if we want to bridge two days and two nights then we would need 6 kWh of cooling energy. Now, using latent cold storage, our original 300 liter tank is four times too big (even after I increased the bridging time). In fact, an 80 liters tank will be sufficient if 60% of the tank is filled with bottles containing tap water. This opens up completely new design possibilities because a 30 cm high box under the fridge will be sufficient.

What worked so well for a fridge might work just as efficiently with air conditioning. We found out that we needed 40 kWh cooling power for the complete night and the tank needed to store that would measure 2.5 cubic meters. If again we filled 2/3 of the volume with bottles containing tap water, then we would gain 154 kWh of cooling power ((2.5 * 2 / 3) * 92.5 = 154.1) so in total we would have 194 kWh available. That is five times more than we need for the whole night; it appears that we need a tank with a mere 500 liters. This tank would be the same size as a sideboard, measuring 90 cm high, 40 cm deep and 1.6 meters long. Not really handy but much easier to put somewhere than a tank holding 2.5

cubic meters (the necessary electric power has to be recalculated because of the lower CoP).

Ice compartment and deep freezer

Water has, by definition, a freezing point at 0°C and when working with water-and-alcohol mixtures the freezing point will depend on the amount of alcohol in the water. Just for your information I supply the corresponding table:

Temp.	-1	-2	-4	-6	-8	-10	-12	-14	-16	-18	-20	-22	-24
Vol.%	3.2	6.3	10.6	16.4	20.4	23.3	26.4	29.1	31.3	33.8	36.1	38.6	40.0

If we replace 20.4% of the water by alcohol in our cooling tank, then we could have a fridge with an ice compartment (within the EU labeled with one star, and the ice compartment temperature is -6°C). If we raise the alcohol content up to 40% then we are easily in the range of deep freezers (temperature -18°C plus/minus 3°C). But we have the problem that we can't use normal water as latent cold storage because that freezes at 0°C.

Our way of getting out of this problem is to use normal table salt. If you happen to live near one of the northern coasts of Europe or America you might know the effect. In winter time all the fresh water ponds are already frozen but not a single sheet of ice can be seen on the salt water of the ocean. That is because salt water starts to freeze at lower temperatures. The correct relationship is pretty complicated but there is a nice rule of thumb: "Sweet water freezes at 0°C and 0% salinity. Salt water freezes at -22°C at 22% salinity." Or even simpler: one per cent more salt and the freezing point goes down one degree.

That gives us a very simple way of adjusting the freezing point to suit our needs. If we want a fridge with a simple ice compartment (at -6°C) then we need water bottles that are filled with 60 grams of salt per liter kept in the cooling liquid (amount of alcohol between 16 and 18%).

If we want to use this system for a freezer then the cooling liquid needs at least 40% alcohol and we must use 210 grams of salt per liter in the water bottles.

I shortly have to explain why not to use salt water as cooling liquid directly. Salt water is corrosive (filled into plastic bottles we don't need to care). Second reason is that with salt water we can't go lower than -22°C but with a alcohol-water mix we can. Thing is that we need a liquid at that temperature in order to distribute the cold/heat.

Modification of normal fridges and freezers

If you are not willing to wait until industry puts such efficient products - suitable for solar feeding - on to the market, why don't you modify your present equipment in a way that will bring some of the benefit? Your aim should be just to bridge the night.

I will assume that you have a normal fridge which is connected to an inverter able to cope with the start-up current (I will come to freezers at the end of this chapter). Then you should have solar panels that are big enough to catch enough power for 24 hours in the day time, a charge controller and a small battery (just big enough to take over when somewhat bigger clouds are passing by). In addition I assume that you have already measured the consumption of the fridge and that you know how much it consumes during the day and also during the night. Lastly, your charge controller needs to be able to switch off the load automatically when the battery voltage goes down too far (else the trick of doing everything automatically will not work).

Your fridge needs to have a separate ice compartment because we are going to use phase shift energy transition between ice and water; the containers of this latent energy (latent because you can't feel any difference in temperature while such a storage delivers or absorbs thermal energy) are simply small plastic bottles filled with water. You will need round about ten liters of water but first read on before you go shopping.

The main trick is to get this water frozen during the day; during the night it will deliver enough cold to keep the fridge cool until a little while after sunrise. That means that the compressor of the fridge is working quite hard during the day and not at all during the night.

Let's assume it is late afternoon, our bottled water is frozen (-2°C) and the sun is going to set. The panels don't deliver enough energy any more and the inverter is pulling energy from the battery. The voltage of the battery goes down and down, and eventually, when it hits the cut-off level that has been set, the solar control unit switches off the inverter (check whether your solar controller can switch the current when the fridge is actually running; if the current is too big you need to place a relay in between). Even if the fridge gives an instruction to the compressor, it can't start to compress.

Because the air surrounding the fridge is warmer than the inside of the fridge, the fridge compartment (6 - 8 °C) loses cold and slowly gets warmer. Now the ice compartment (-2°C) has to deliver cold to the fridge compartment. After a while we have a nice cool 0°C in the ice compartment and the ice in the bottles is 'activated'. The conversion of ice into water starts and the bottles keep their temperature of 0°C as long as there is some ice left within them (and that can take quite a few hours and that is exactly the purpose of the exercise).

This means that your fridge motor (and compression pump) would want to start since the temperature is about 2°C too high (actually, the motor can't since the inverter is switched off). If the amount of bottled ice in the ice compartment was big enough the temperature will stay until the next morning.

The next morning, shortly after sunrise, the charge controller starts to fill the battery; after a while the voltage is high enough that the inverter is switched on again. The fridge will sense that the temperature in the ice compartment is 0°C (we hope) instead of the -2°C we had selected with our thermostat and the compressor is started. A new cycle starts with cooling the content of the fridge and converting the water in the ice compartment into ice once more.

Now the question is, how much water/ice do we need? Here we make use of the measurement you made of how much electricity the fridge used at night. Since I can't know how much your fridge uses I just assume that your fridge used 0.5 kWh per 24 hours and that 0.3 kWh of that was needed during the hours without sunshine. With a modern fridge the CoP (the heat pump multiplier effect) should be somewhere near to 3, so you will need about 0.9 kWh of thermal energy for the night.

Water has a freezing enthalpy of 333 kJ/kg, which corresponds to 0.0925 kWh/kg; that is near enough 0.1 kWh/liter. So in order to provide the 0.9 kWh of cold during the night we will need to freeze 9 liters of water.

Because the cold needs to be delivered from the ice compartment to the rest of the fridge we need a stream of air. If the 9 liters were a solid block in the ice compartment there is no chance that all the thermal energy could be delivered. So what we need are many small blocks (like thermal packs) or small bottles with a bit of space between the blocks or the bottles. That ensures a very big surface, short distances for thermal energy transport and good ventilation.

If you can't manage to squeeze 9 liters into the ice compartment you will have to set the cut-off voltage of the solar control unit so that the battery delivers more energy in the evening before the bottles with ice have to take over (to use salt water in the bottles does not help since you only change the freezing point but not the amount of stored cold). There is no other way and you will have to remember: no warm beer bottles into the fridge in the evening (unless there is an emergency)!

Suppose that you want to use this economic system in a freezer. A deep freezer should be set to -18°C and the actual temperature must not differ more than 3 degrees (at least that is the rule for freezers used in German gastronomy and retail trade). Let's assume that you adjusted your freezer to -18°C. Now you fill your bottles with water containing 160 grams of salt per liter and you put them somewhere in the freezer (it is best to put them somewhere near the top; and again you should use many small bottles instead of a few big ones).

Again the amount of water needed is calculated using the amount of electricity consumption you measured during the night. That we moved the freezing point by help of salt does not change the enthalpy (at least not enough for it to be important). So again 1 liter of ice can store 0.1 kWh of cold. You multiply the electric consumption during the night with 30 (CoP of 3 and 10 liters per kWh) and you get the necessary liters for the latent cold storage.

The connection to the solar installation can be exactly the same as with the fridge. So here is what will happen. The temperature in the freezer might oscillate between -17°C and -19°C during the day. This should be enough to convert all the salt water in the bottles into ice (check it!). Then the sun sets and the temperature within the freezer goes up. If you calculated the number of bottles (liters) correctly the temperature within the freezer will still be at -16°C in the morning when the sun starts to shine.

In fact what you are doing is to change the wear and tear of the battery against some space in your fridge/freezer. Make up your numbers in order to find out whether this is a paying proposition, whether the savings you make by buying a smaller battery are equal or more than the need to spend more on a bigger deep freeze.

Apropos pay for itself. In order to finish this subject I have a short look at the economic aspect. A latent cold storage costs near to nothing and one will only use it if there is enough free space available anyhow. The investment costs are practically zero. In the solution where we switched over to mains during the night we had a consumption of 0.3 kWh during the night costing in Spain 6.6 cents and we are saving this money now. During 20 years we would save 482 Euros; enough money to buy a new fridge.

7. Supplement: batteries

No previous knowledge is necessary for understanding this chapter and there is not a single new formula; we are only talking about simple interrelated facts. In fact, I'm quite sure that you never looked at batteries in this way before.

If you are planing a photovoltaic installation which feeds directly into the mains it is not necessary to think about batteries. The same applies if you have a direct connection to the grid. Everybody else has to think about batteries, or else it will be dark in the house at night.

General information

There are a lot of different types of batteries and they come in different sizes; and the number of possible applications is endless. This chapter is limited to the use of normal lead acid batteries in photovoltaic installations since they are still the cheapest way of storing electric energy for intermediate lengths of time and they only need a fairly limited amount of care and attention. If you are thinking of using VRLA (valve regulated lead acid) batteries or gel batteries, then you should read this chapter carefully, which may make you change your decision.

The main source of information for this chapter is the doctoral dissertation of Dirk Uwe Sauer (Ulm/Germany 2003). In his dissertation Sauer developed a model not just for simulating batteries but for battery management as well. His work forms the basis of this chapter and I have added some bits and pieces. Nearly all the graphics in this chapter were copied from his dissertation and are used by courtesy of professor Sauer.

At first sight it would seem that lead-acid batteries are pretty simple devices, but if you think so, you are utterly wrong. Though this battery type was invented more than 100 years ago, it was not until the dissertation of Dirk Uwe Sauer that the first complete simulation of this battery type was published (well, the simulation was so complex that the computer program was at times only slightly faster than the actual processes taking place in the real battery).

In this chapter you will find a simplified description of how such a lead acid battery works, what additional (side) effects might take place and how these influence how long the battery actually lasts. Bad maintenance and management of the battery can have serious financial consequences. What you will learn in this chapter will not make you a battery expert but will enable you to judge whether anybody is trying to sell you some bull . . . . about batteries. It will also tell you some of the more important things that you must do or not do - all other mistakes will be minor and relatively cheap, so don't worry about them. If you want to avoid these minor problems as well you will have to get yourself a good book on this topic such as those used in technical colleges.

All batteries have components that wear and tear. Depending on what treatment they get, they will last less than a day or up to more than 20 years; in domestic PV installations using 'solar batteries' we can expect 7 years if given good treatment; when using real traction batteries the time could be much longer. All batteries, however carefully looked after, will die a slow death even if not stressed, but if they are stressed they die sooner, and if mistreated they will break down very fast. Typically batteries in a photovoltaic system account for 20 to 50% of the total cost over the period of 20 years that a stand-alone PV installation is expected to last, so it does make sense to spend some time thinking about them.

Depending on their application, the design of the inner structures of the battery may show significant differences. In an UPS (the uninterrupted power supply used in telecommunication centers when mains electricity fails) the battery is stressed slightly once a month during the normal system check and is really used only once every few years for a couple of hours. This type of battery is stored and maintained under absolutely perfect conditions and it can easily last 20 years.

Traction batteries, like the ones used in fork lift trucks, are stressed heavily but work under controlled conditions at all times. Their capacity is such that they are good for a complete shift (typically 8 hours), after which they are fully charged (the remaining 16 hours). Everything is avoided which might stress the battery unnecessarily, and this is also what we aim for in domestic PV installations.

Batteries in stand-alone solar installations can only be charged when there is sufficient solar radiation. How much and how often the batteries are discharged depends on the actual situation, and when they can be charged again depends on the weather. First the demands of the user have to be met and only then comes the need to handle the battery with care. It is obvious that costs increase progressively with the demands of the user!

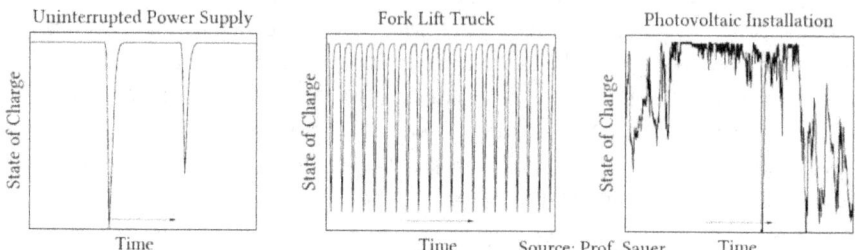

Source: Prof. Sauer

How does a lead-acid battery actually work?

All batteries have two connections, called the positive and the negative pole. Between these poles electrical tension can be measured, given in Volts. The pressure of car starter batteries will normally be 12 Volts and for the batteries in the bigger commercial vehicles 24 Volts. If an electrical device such as a light bulb is connected between these two poles, this electrical tension pulls the electrons from one pole through the device to the other pole of the battery; in order for a current to flow you must have a closed loop. The number of electrons being pulled through this loop per second is called electric current and is measured in Amperes.

In earlier days the light bulbs in cars were all of the same base type; an evacuated bulb with a filament of tungsten inside. Some of these light bulbs burned very bright, like the ones in the headlights, other bulbs were less bright like the bulbs in the brake lights and other were relatively dim as the light over the number plate. These bulbs have different power consumptions depending on the brightness. The power is given in Watts and with DC current it is simply calculated by multiplying current by tension (Amps * Volts). If you connect a 6 Watt bulb to a 12 Volt battery there will be a current of half an Ampere.

When such a device is connected to a battery, a chemical process starts delivering electrons to one of the electrodes. From this electrode the electrons flow, via the pole, through the device (in which they perform some work) and then via the second pole, to the other electrode. There they must be transformed by a second chemical process. In the first chemical process not only are electrons set free but also ions (ions are atoms or molecules which miss one or more electrons). The electrons go along the wire and through the electrical device, while the ions flow through the battery liquid to the other electrode. Here, in the second chemical process, ions and electrons are recombined. The power pushing the electrons through the wire is the pressure between the electrodes, measured in Volts.

Every battery has therefore three active parts: a positive electrode, a negative electrode and the battery liquid. In a fully charged battery the negative electrode consists of pure lead (Pb), the positive electrode consists of pure lead dioxide (PbO_2) and the battery liquid is diluted sulfur acid, that is (H_2SO_4) plus water (H_2O).

When discharging the battery, lead sulphate ($PbSO_4$) builds up on both the positive and the negative electrode. When the sulfur combines with the lead, the battery liquid in which the two electrodes sit is diluted. Since lead sulphate does not dissolve easily, the newly generated lead sulphate molecules will start looking for seed crystals so as to connect to them. The combination of the lead with the sulfur is a chemical process that happens directly on the surface of the electrode (the electrons need to go where the action is) so the probability is very high that the seed crystal will be found on this surface of the electrode or at least very near by.

When a battery is charged electrons are pushed by an external pressure in the opposite direction. On both electrodes the lead sulphate is dissolved; the sulfur goes back into the battery liquid and the lead and lead dioxide is deposited on the positive and negative electrodes again. Ignoring heat losses, the processes that take place when discharging the battery can be reversed by charging the battery. In theory one should be able to charge and discharge lead acid batteries as often as required; we will see in a moment that in practice this is not possible.

With reversible chemical processes, in one direction the conversion always produces heat but in the reverse process takes up heat. So even with a perfect battery there will be losses. With a real battery there are additionally the losses produced by the current when flowing through the different materials - the electrons collide with the atoms and molecules of the material and these collisions produce heat which is lost to the surroundings; heat is ultimately the kinetic energy of atoms and molecules.

However, the biggest energy losses in a lead-acid battery are caused by electrolysis, the process whereby the water in the battery liquid is broken down by the electrical tension between the two electrodes to produce oxygen and hydrogen. The tension between the electrodes is, even with nearly flat batteries, still high enough to cause this decomposition (it only need 1.9 Volts). The losses by gassing has been estimated as high as 95% by some authorities. Electrolysis is the reason for self-discharge.

Electrolysis uses quite a lot of energy and the amount of oxygen and hydrogen gas produced on the electrodes depend very much on the voltage. A nearly flat battery will only produce very small amounts and a nearly full battery will produce quite a lot, especially when it is being charged at a voltage that is slightly too high. Because of the oxygen and hydrogen gas produced, batteries have to be kept in well ventilated places, especially when being charged.

Oxygen and hydrogen by themselves are harmless but the mixture produced by the batteries is explosive. When examining the statistics of cars ruined by fire because of exploding batteries, one will know that the danger is not all that great, but even so, remember: no open fire or sparks near batteries!

General construction of a lead-acid-battery

There are many ways that lead acid batteries can be used. Their use as a starter battery in vehicles is known to everybody; this is where a lot of power is needed for a short period of time just to get the main engine of the vehicle turning. Another widely used application is as a battery backup power supply for technical installations, in which most of the energy is needed during the regular system check-ups and for the rest of its 20-year lifetime the battery is hardly ever stressed. The types of battery that are stressed most are what are usually called traction batteries, which get used in fork-lift trucks, golf carts and electric wheelchairs. These batteries usually get discharged and charged once per day. Depending on the use to which the battery is put, their inner structure has to be different.

The price of a battery in the shop depends ultimately on the amount of materials used plus the work required to form the materials in the desired way, such as producing the spongy lead foam used in the electrodes. The price also depends very much on the number of units being produced, which is why starter batteries for cars are relatively cheap. Traction batteries are produced in fairly large numbers as well, but solar batteries are still a niche market, which is why they are quite a lot more expensive. In fact, since solar batteries don't do anything that a normal traction battery couldn't do just as well, their price is determined by the fact that they are marketed as a sort of lifestyle gadget.

Ideally we want a battery which can store a lot of energy in a small volume with little weight and even smaller costs. Additionally it should be powerful, in the sense that the stored power can be consumed in a short time whenever needed. So we start with the simplest battery possible: we have a trough filled with sulfur acid, a massive plate of lead for one electrode and a massive plate of lead dioxide for the other electrode (this is by the way the definition of a fully charged battery: a lead electrode, a lead dioxide electrode, diluted sulfur acid and not a trace of lead sulphate). Such a battery would work and deliver the expected voltage. But how much energy could it store?

When we draw energy from this battery, the surface of both electrodes will be covered with a very fine layer of very small lead sulphate crystals which are separate at first but which will rapidly grow towards each other. After a relatively short time the complete surface of each electrode will be covered by badly conducting lead sulphate crystals and the power that the battery can produce will go down rapidly. Each single chemical action will yield (or consume) one electron. Since the voltage of the battery stays (more or less) constant, the amount of delivered power is directly proportional to the amount of lead sulphate deposited. If one wants to deliver more energy one will simply have to provide a bigger surface!

Over the last decades the design of lead-acid batteries has been developed further and further; current technology uses electrodes which have a skeleton made from a very good conducting material; a paste with a high content of water is placed on it and when the water evaporates it leaves a very spongy material; the positive electrode consisting of lead dioxide that has a chemically active surface of 5 m^2 per gram and the lead foam of the negative electrode comes up to 0.5 m² per gram. So the electrodes consist mainly of a system of many small and interconnected caves filled with sulfuric acid and even a medium sized starter battery will have electrodes that have a few thousand square meters of chemically active surface.

When such a battery is discharged, the walls of this cave system get covered with a thin layer of lead sulphate (many small crystals) and as the battery is discharged, more and more of the active material of the electrode is converted to lead sulphate. Lead sulphate molecules occupy a larger volume than pure lead or lead dioxide (the volume ratio of $PbSO_4$ to PbO_2 is 1.94 : 1; that of $PbSO_4$ to Pb is 2.40 : 1). The caves in the electrode become more and more clogged up as discharging continues. The sulfuric acid is chemically diluted and less and less fresh acid can flow into the channels connecting the caves; the battery is ailing but can recover with time (by the process of diffusion).

At this stage we arrive at the point in time where the process starts which is responsible for the destruction of batteries by deep discharge. The lead sulphate starts to fill up the pores of the electrode completely. Any additional molecule will apply actual mechanical stress to the walls of the caves. The material of the electrodes is not very elastic, so that cracks will develop in the walls between the caves. If the material were not stored in what are usually called "bags" (plastic envelopes with lots of little holes) the electrodes would crumble away and sink down to the bottom of the battery case. Death by micro-mechanical destruction!

In fact, a single deep discharge can damage a battery so badly that it is not fit for normal use any more! Depending on the type of design (and the porosity of the electrodes) there are different kinds of robustness. Starter batteries can deliver a lot of power from a small volume; their electrodes are extremely porous. Batteries for forklifts don't need to deliver very high peak powers and they can be bigger and their electrodes are much more robust.

That means that starter batteries are the worst choice for PV installations (unless somebody gives them to you and you only have to take responsibility for their correct disposal) since the internal destruction which occurs every time the battery is charged and discharged is considerable. There is no way to force the crystal growth to take place only in places where it is wanted. Therefore there is no way to suppress this micro-mechanical destruction.

In a PV installation one wants to discharge and charge the battery as often as possible. Traction batteries are designed for many cycles and they are produced in larger quantities, which is why they are relatively cheap (a bit more than one Euro per Ah at 12 Volts in 2013; look for forklift batteries) and most producers have their own delivery service (that means no problems with carrier companies declining to handle delivery because of the acid).

Another way to solve this delivery problem is to order the batteries delivered in a 'dry' state, that is, there is no acid in the battery (the acid comes in separate sealed plastic bottles) so that they are not seen as hazardous goods. The definition of a fully charged battery is that it does not contain any lead sulphate at all. In a 'dry delivered' battery (don't confuse such a battery with the dry battery in your torch or radio) there is no lead sulphate, so that when the battery is filled with diluted sulfuric acid it becomes fully charged. In actual practice the SOC (State Of Charge) will be only 60 to 80%; after filling the battery with acid - you should charge it properly as soon as possible.

If you buy a ready-filled battery, make sure that you know the production date or rather, more specifically, the date the battery was filled with its electrolyte. Think twice before buying a battery that is older than half a year. The reason is that lead acid batteries will self discharge, usually (unless stated differently) at the rate of about 10% per month. After half a year the battery might be severely damaged; after a year it will not only be flat but dead as well.

A few definitions

Since batteries are an article of every day life, the meaning of the technical terms used in talking about them may differ considerably from their generally used meaning. So we need to give some definitions.

A battery is really fully charged when there is no lead sulphate whatever on the electrodes. This condition can only be reached with new batteries and only after a few days of charging. 99,99% SOC (State Of Charge) is the scientific interpretation of "fully charged". When viewed from this point of view, car drivers are utterly wrong when they think that twenty minutes driving with high rpm will fully charge the battery; although in daily practical terms they are reasonably correct. How best to charge a battery will be discussed later.

The nominal capacity of a battery is given in Ah (Ampere hours). A battery with 100 Ah should theoretically be able to deliver a current of one Ampere for a period of 100 hours. Some capacity remains but should not be used because that will cause destruction.

Often one finds a statement such as "The I_{10} of this battery is 9.5 Amps" by which is meant that a fully charged battery will be discharged within 10 hours if a current of 9.5 Amps flows (from any one battery type to any other battery type the actual I_{10} delivered could be different). With a statement such as "This battery has a C_{10} of 128 Ah" the capacity of the battery is being referred to if that battery is discharged over 10 hours. These kind of statements are necessary since any given battery has a higher capacity with a low rate of power consumption than with a high rate of power consumption (I told you batteries are no easy subject!).

The amount of energy which can be stored in a battery can (roughly) be calculated by multiplying the nominal capacity by the nominal voltage. A battery with 100 Ah and 12 Volt nominal tension can store 1,200 Wh or 1.2 kWh. In other words, a typical starter battery can store electric energy worth about 30 Cents if that energy had come from the grid.

A battery is defined as not being operational any more if it is down to 80% of its nominal capacity (DIN 43539 / part 4). If a battery is down to 75% a technician would want to replace it. The owner of the solar installation would still complain years later what a waste of money this would have been, since the battery had worked perfectly for another three years! This is the main reason why opinions about successful reanimations of ailing batteries differ so much.

Crystals: how they grow and how they dissolve

In order to be able to understand some of the processes taking place in batteries we need to take a closer look at crystals and how they grow. Crystals are formed from salts dissolved in liquids; the salts exist in both the liquid and the crystal form (for example ordinary table salt and water). When the concentration of the salt in the liquid is high (and if what are known as seed crystals are available) there will be the growth of crystals, which therefore lowers the concentration of salt in the liquid.

Thus there is at all times a balance: if the concentration of salts in the liquid is high the crystals grow and will lower the concentration; if the concentration in the solution is low, the crystals will dissolve and the concentration in the solution will rise. Maximum possible concentration depends on temperature: the higher the temperature the higher the possible concentration (as a rough rule).

Now we need to take a very close look at such a crystal. The building blocks of a crystal are held together by means of a physical force, called the bonding force. At first sight there is no reason at all why such a building block should leave the crystal and go into solution again (because of the existence of this bonding force, it takes energy to separate the building block from the crystal). Now we introduce the idea of heat, which is the common name for kinetic energy, which refers to the speed of atoms and molecules moving about in a gas, a liquid or a solid.

Think of a very small crystal consisting of just two building blocks; they are globes just touching at one point. Thermal blows - atoms and molecules moving in the battery liquid - which could separate the two globes could come from nearly every direction, they only have to be hard enough to cause the building block to overcome the bonding force and leave the crystal. Now think of a crystal consisting of a few thousand building blocks, neatly packed together in a globe. The building blocks inside the globe are complete safe and can't be forced to leave the crystal and dissolve directly. The building blocks on the surface are protected against thermal blows from below and from any side. Blows coming from above will consolidate the structure but will not harm the crystal. A single blow will normally not be enough to separate a building block from the other building blocks in the crystal. In fact several blows are needed to get the crystal vibrating so hard locally that a building block might be ejected. This is not going to happen too often.

If we have a small crystal next to a large crystal the probability that a building block will leave the small crystal and dissolve will be much higher than for a building block from the larger crystal. Therefore, with the concentration of the solution staying constant, large crystals will generally grow at the expense of small crystals. The amount of energy necessary to knock a building block away from the crystal, i.e. overcome its binding force, stays constant, but with rising temperatures this amount of energy is available more often. The warmer the solution in which it sits, and the smaller a crystal is compared with its neighbor, the faster it will disappear.

Of all bodies, a globe has the smallest surface area in relation to its volume. Corners and edges in an irregularly formed crystal have the same effect as small crystals. If there are crystals in competition (this is even true for different surfaces of the same crystal) they will tend to form as big a globe as possible (though there are crystals which do not form globes). When charging a battery the number of sulphate crystals going into solution depends on the concentration of the battery liquid and the size of the reachable surface. That is, the percentile loss of small crystals is much bigger than the loss of larger crystals. Large crystals resist losing their building blocks, which is why it will take much longer times to dissolve them.

Sulfating

Building up sulphate crystals forms the very basis of how lead-acid batteries function. The crystals build up when the battery is discharging and then dissolve when the battery is charging again. Sulfating takes place at all times when the battery is discharging. But mostly the term "sulfating" is used in the sense that crystals built up that are so big that they do not dissolve during regular charging. We need to have a closer look at that.

Let's assume we have a brand new battery and there are no sulphate crystals in it yet. Now we connect a device that uses a lot of energy between the positive and the negative poles, and measure the battery voltage very accurately and with high

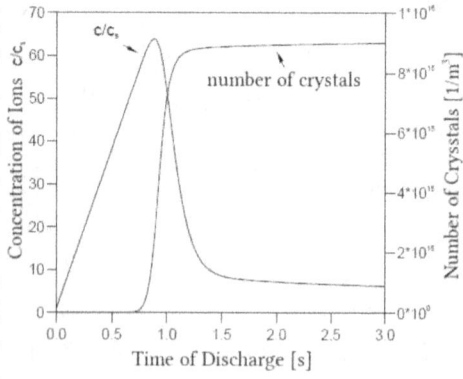

resolution. We would expect to see the voltage drop to a certain level and then to go down very slowly (more or less at a constant rate) while the battery discharges. In fact, the voltage drops for roughly 1.5 seconds far under the expected value but recovers again and then starts to drop with a very low and constant rate.

This effect can be explained by the fact that immediately after connecting the device there were no seed crystals as yet, and therefore the concentration of lead sulphate in the battery liquid increased rapidly, far above the normal balanced level. It has been estimated that 4 to 8 molecules of lead sulphate must come together in order to form a seed crystal. But such a mini-crystal can't just float in the solution, it needs something to which it can attach itself to or it will go into solution again. Since the chemical processes happen directly on the surface of the electrodes, the concentration there will be highest and the surface of the electrodes will be the perfect place for the seed crystals to settle down. Within a very short time there will be a gigantic number of small sulphate crystals on the electrodes and the concentration in the solution drops significantly. Later the driving force for building new seed crystals does not exist (the concentration is too low), so we are from then on in the zone of crystal growth and shrinking.

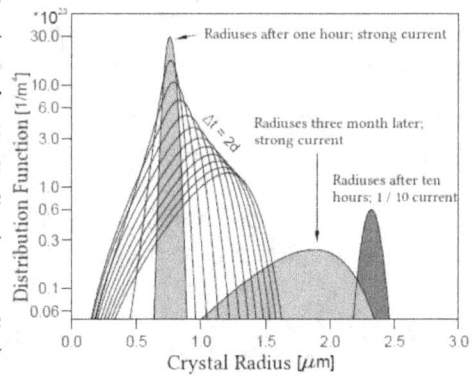

Generally it is said that there is hardly anything worse that can happen to a battery than to leave it half discharged over a longer period of time. But how bad it is depends on the history of the battery. In fact, batteries do have a memory, which is represented by the number of seed crystals.

In the picture above we see the result of a simulation. A battery was discharged, using a high current, down to 50% DOD in the space of one hour and we see the distribution of crystal sizes directly afterwards (first gray peak). There are many crystals and they all are more or less the same size. Afterwards the battery was left alone and there was a competition between the crystals. Those that were a bit bigger started to grow at the expenses of the slightly smaller ones. So the number of crystals dropped rapidly (there is a logarithmic scale on the number axis) and we get very small crystals and bigger ones. Then the very small crystals disappear and the distribution of sizes gets broader. The lines of the array of curves were computed for every other day. The broad light gray peak gives the distribution of sizes after three months. The process of redistribution slowed down significantly.

These curves can't have a clear meaning to us at the moment. This meaning comes when we compare the distribution of sizes after 3 months with another distribution. The dark gray distribution we would get if we discharged a full battery over ten hours down to 50% DOD. Immediately afterwards the distribution of crystal sizes is much worse (the crystals are bigger) than with the other distribution after three months.

If we have a fully charged battery (no lead sulphate on the electrodes) we normally can not know what will happen in the future. So it is best to discharge the battery during a few seconds with a really strong current (it costs about 0.3% of the precious energy if we discharge the battery during 3 to 5 seconds with the maximum permitted current). This makes sure that the maximum number of seed crystals is generated and this will slow down the aging of the battery by sulfating. Additionally many small crystals mean that a really large surface is created, and the bigger the surface the greater the power of the battery.

When charging the battery, many small crystals mean a big active surface and the possibility of dissolving the sulphate crystals nearly completely in a short time (we want to keep the seed crystals alive). The charging phase with a steady current (see later) can be longer and the time necessary for a full charge will be shorter.

Charging a battery and lead sulphate

Let's take a look at the positive electrode. To charge a battery we need to push electrons into it (in practical, every day terms: we have to connect the battery to a power source with a higher voltage than the actual open circuit voltage of the battery). The electrode consists of pure lead and is covered by lead sulphate crystals. Lead sulphate is a bad conductor so that electrons will find it difficult to pass through this covering coat of lead sulphate crystals. But without electrons we can't convert the lead sulphate into lead and sulfur acid. Therefore the best candidates for conversion are those parts of the crystals which are in direct or almost direct contact with both the pure lead and the electrolyte.

In order to charge the battery, lead sulphate molecules have to be dissolved in the battery liquid, which they are unwilling to do. Depending on the temperature there is a certain concentration of lead sulfur molecules in the battery liquid (the higher the temperature the higher the concentration and the diffusion speed). When such a lead sulphate molecule hits by chance the pure lead of the electrode when there is a free electron available at the same spot, conversion will take place (a lead atom connects with the lead electrode and a sulfuric acid molecule is produced).

The longer the average way for a lead sulphate molecule is, the slower the charging process will be (well, one can help a bit with higher voltages). So it is logical that first the sulphate molecules at the edges of the crystals are reduced.

The lead sulphate in the direct neighborhood of the lead of the electrode will vanish instantly. Let's assume we have a situation where we have a surface of pure lead next to a sulphate crystal at a distance of about a molecules size. The top layer of the sulphate crystal will disappear rapidly and, since a lead atom is only 40% of the size of a sulphate molecule, the distance to the next layer will be considerably larger. When there are lots and lots of small crystals this effect will be fairly fast, but if there are thick crystals the distance will increase more and more and the sulphate molecules have to traverse a longer distance by diffusion. Charging will become slower.

When a battery is discharged the sulfuric acid combines with the lead and the lead dioxide. This lowers the specific gravity of the acid. With a new and fully charged battery the specific gravity will be highest and it drops down in proportion to the energy delivered by the battery. Charging and discharging of a new battery can be measured directly by measuring the specific gravity of the acid.

Because of the voltage between the electrodes (and the corresponding poles), water will be split into hydrogen and oxygen. These gases will escape and diffuse. That means the level of the electrolyte will drop and the specific gravity of the acid will rise. The charging state of a battery which has lost water by gassing can not be checked by measuring the specific gravity (first the electrolyte has to be filled up with distilled water which must mix with the battery liquid).

If a battery is older (or has suffered because of poor maintenance) there will be bigger sulphate crystals on the electrodes and they will not even disappear when the battery is charged over a longer period of time (for instance, a few days). The concentration of the acid can not reach the levels of a new battery any more. In such situations now and again people come up with the idea of just adding more sulfuric acid. In principle this is not wrong but the problem is that the amounts of active materials in the battery are adapted to each other. Filling in more acid does not increase the amount of lead or lead dioxide. Filling up the battery with acid is therefore pretty useless!

In a new battery the mass ratios of the active materials (lead, lead dioxide, sulfuric acid) are near their optimum. Since one can not know how much a battery already has suffered, the state of charge can not be determined by the specific gravity of the acid. On the other hand one can connect such a battery for a couple of days to a good battery charger and measure the specific gravity then. Comparing this value with the value of a new battery will give a good idea about the state of health (SOH). If you want to know it more precisely you will have to measure the battery under load.

Acid stratification and the effects

In order to be able to compare his theoretical calculations concerning acid stratification with the real life situation, Dirk Uwe Sauer dismantled three normal batteries for his dissertation and put the working parts (i.e. the electrodes) into a deep trough, one above the next. These three batteries were interconnected, outside the trough, so that it was possible to measure the currents for the sections separately (top, middle and bottom) and to know the individual SOC (State Of Charge) of each of them. In the actual experiment, two full cycles (discharge and then charge) were followed by three part cycles (which is not untypical for PV installations) followed by a full charge. It is obvious from the chart that the state of charge of the three batteries drifts apart significantly during the time of charging.

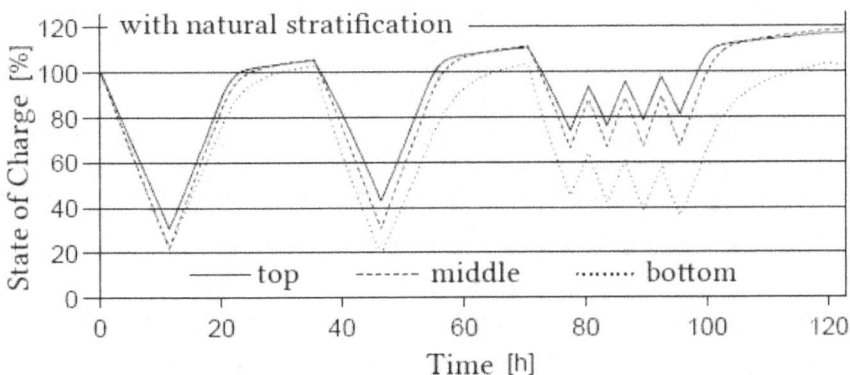

If such a battery were left alone after the third part cycle then the top part would be at 80% SOC and the bottom part would only have a SOC of 40%. The best conditions for crystal growth would actually be in the bottom set of electrodes. The battery would die fast, starting from the bottom.

But it gets worse and this has to do with the amount of active material that is available. Let's assume that we could collect most of the sulfuric acid from the top of the trough and transport the acid to the lower part of the battery. This would be some kind of extreme stratification.

When we now discharge the battery, the remaining sulfuric acid in the top part is used up relatively fast which means that this part of the battery is fully discharged (as far as the electrodes are concerned more would be possible but there is no acid left). Therefore there will be nearly no deterioration in this part of the battery. In the lower part of the battery we have plenty of acid so the discharge of this part can go on much longer. The porous material of the electrode fills up more and more with lead sulphate (remember the different volume ratios) and the electrodes will suffer mechanical strain as if they were going through a deep cycle that is too deep. The battery as a whole would seem to be in reasonable condition but the lower part would be in mortal peril.

If the battery is charged afterwards then the little bit of sulphate in the top part of the battery is converted very fast, because the crystals are all very small. Already after a short time the upper part of the battery is fully charged (per definition since there is no sulphate left) but the lower part is still being charged.

In summary the top tier of the battery is going through small cycles (of charging and discharging) and is likely to have a long and healthy life. But this is not true for the lower part. Even if the battery as a whole only discharges to 60% the lower part does full cycles deep into the danger area. These are the best conditions to develop thick sulphate crystals covering the electrodes, lowering the capacity and destroying the structure.

If we had batteries consisting of different parts (like in the experiment) then it would be easy just to replace the lowest part when it is used up. But with a normal battery this is not possible. Lets assume the top 2/3 of the battery are still in perfect working condition but the lowest part is down to 40% SOH (state of health). By definition a battery is not fit for use any more when its actual capacity is lower than 80% of the nominal capacity. The stratification can easily cause situations where batteries only last 10% of the predicted time! And you did nothing obviously wrong.

Traction batteries are often relatively high (like three 'normal' batteries on top of each other) so one can really expect pronounced stratifications. Those batteries which seem to be the most economic solution are easiest to destroy.

Before thinking about how to fight stratification we must have a look at how it develops. When a battery is charged lead sulphate crystals dissolve and the molecules are split into lead, lead dioxide and sulfur acid. Since this happens within the pores of the electrodes, high acid concentration builds up there (it is precisely where the chemical action takes place). The acid inside the caves is heavier by far than the acid outside between the electrodes (1.84 gr/l to less than 1.20 gr/l). That is why this acid is going to leak down through the porous material of the electrodes. If gel-batteries are used this will not happen since the gel prohibits this type of flow. In VRLA-batteries a fleece soaked with acid separates the electrodes; even if the fleece was hindering any flow, the porous material of the electrodes is not. So stratification in VRLA-batteries takes place and there is no way of reversing it. The Why we will see right now.

So what to do against stratification? In early days gassing was the only available solution. Every few weeks, after a normal charge, the voltage was raised a bit more and water was split by electrolysis. Gas bubbles were generated and these bubbles mixed the electrolyte while rising to the surface. But the quality of this mixing was insufficient at the bottom (few and small bubbles) and good near the top (many bubbles and quite big ones since while rising the bubbles grow in size because of the lower pressure). For good results a long gassing time was necessary.

This method is much better than nothing but it has three disadvantages. The first is that it needs a lot of energy. The second is that is uses up the water in the battery (distilled water is not really expensive but the time for regular maintenance is, at least in industrial applications). The third (but small) problem is that the bubbles are generated directly on the surface of the electrodes and when they rise through the acid they jolt the porous material.

A gel-battery uses a sealed case. If gassing occurs in such a battery, the gases have to recombine to form water. If the amount of gas is small there is no problem. If the amount is high the gas will escape through an emergency vent. The same happens in AGM-batteries, which is why they have a second name (VRLA = vent regulated lead acid battery; gel-batteries do have such a vent as well but they are seldom called VRLA). If the pressure gets too high the gas just escapes. But later the equivalent amount of water will be missing and there is no way of replacing it (no stopper). Therefore you should never connect gel- or VRLA-batteries to battery chargers with the gassing (or equalizing) option switched on.

We saw that already two full cycles plus a few part cycles were sufficient to produce a pronounced stratification. Under such conditions a gassing phase every two to four weeks (as recommended by many people) will be nothing but applying decorative cosmetics. Owners of PV installations are normally happy when they get their batteries charged at all. During summer time it might be possible to get the batteries gassing frequently, but in winter time, when charge/discharge cycles are as deep as necessary, there will be neither the time nor the possibility. However, there is a solution that is little known in private applications: electrolyte circulation!

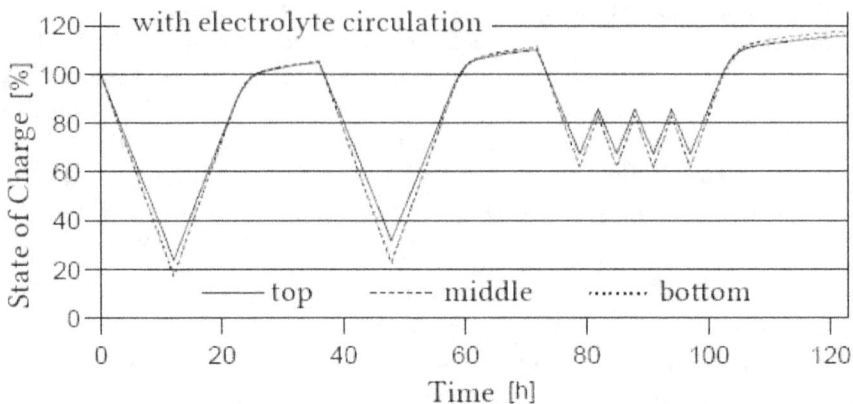

The term electrolyte circulation could be used for gassing as well but it is normally restricted to applications using a pump. The effect of such a pump can be seen in the chart. Since there are no differences in acid concentration the differences in capacity are gone as well. All parts of the battery are aging with the same speed.

Normally it will not be possible to use a submerged pump for electrolyte circulation - at least I could not find any manufacturer offering batteries with submerged pumps. What I did find are batteries with a predetermined breaking point (there are small circles on the top of the cell indicating where you can make a hole of roughly 8 mm). Here you can insert a thin tube going down to near the bottom of the battery. This tube you connect to an air compressor. The compressed air goes down into the battery and air bubbles go up and by friction they pull up the dense acid to the surface. In effect the stratification is eliminated. Experiments seem to show that a mere three minutes of pumping per hour are sufficient.

For the compressor we need an air pump of the type used in aquariums. Generally these pumps are proof against the acid used in batteries. Pumps of this kind start at about 2 Watts and they are capable of pumping air down into water or acid to a depth of half a meter. The capacity of such a pump will be good for several cells.

The electrolyte circulation does not only prevent stratification but also guarantees an even heat distribution (the electric current in the electrodes heats them up and the flow of acid transports the heat to the case surface). This optimizes the heat dissipation which in turn increases the life time of the battery.

Such electrolyte circulation is only necessary while charging the battery. This is because only then concentrated sulfur acid is generated on the surface of the electrodes. In practical terms, the need for electrolyte circulations happens in a PV installation exactly when most energy is available.

It is very difficult to make an accurate estimate of the financial savings due to electrolyte circulation. Comparatively little work is needed to build a small air lift pump and the price of the tubes and the pump itself are just a few Euros. Since the electrolyte circulation is the best you could possibly do for your battery it should be worth a try.

Since batteries react badly to introduced dirt there should be a small filter in front of the air pump; this filter must be so fine that no dust can pass through.

If you use one air pump for two (or more) cells the very small differences in how deep the tubes go down into the battery might cause somewhat bigger differences in the amount of air going through the cell (if, for example, the battery is slightly tilted). I found a good tip in a forum about photovoltaics; you use a normal plastic tube as a stethoscope. One end you put into your ear and the other end you hold over the hole for the stopper. You will be able to hear the difference between adjacent cells; in those cells that do not bubble loud enough you just lift the venting tube slightly (or if one cell is much louder you stick that tube deeper in).

Corrosion in a battery

If you hear the word corrosion in connection with lead acid batteries you will almost certainly think that the acid has something to do with it. That is pardonable, but in fact utterly wrong. The reason for corrosion is based in the electrolysis and it works like this: starting at 1.9 Volts the water starts to dissolve (the tension is high enough to separate the oxygen and hydrogen atoms from each other; in order to overcome the binding forces quite a lot of energy is necessary). This separation can take place anywhere between the electrodes. The hydrogen starts to move to the negative electrode and the oxygen is moving to the positive electrode.

Hydrogen and lead dioxide don't see any possibility to react chemically with each other. Therefore hydrogen builds up little bubbles, then bigger bubbles and they start to rise in the liquid and disappear from the battery. The oxygen moves to the positive electrode and encounters lead. Here we can have easily a chemical reaction since oxygen and lead can (and want to) combine to form lead dioxide.

Since any cell in a lead acid battery has a nominal voltage of 2 Volts, which is definitely above 1.9 Volts, there is no way to stop the electrolysis completely. On the other hand, if the voltage were lowered that far then one would have big lead sulphate crystals in any required amount. So again we have the choice between the devil and the deep blue sea. Therefore the manufacturers of batteries decided to put more pure lead into the electrodes so to postpone battery death a bit.

Now we are back in the cycle. Charging the battery produces concentrated sulfuric acid. Sulfuric acid is the reason for stratification. The stratification can only be removed by heavily gassing. Heavily gassing is the best way of generating oxygen which in turn starts the corrosion.

It looks as if there were no way out but there is and we already know it: electrolyte circulation! Then we don't have to force the gassing for the electrolyte circulation and we reduce the production of oxygen drastically. That means in turn drastically reduced corrosion!

So there is no help but to put pressure on the battery manufacturers so that they start to offer systems for electrolyte circulation at acceptable prices (or at least that they produce batteries in which it is easy to install such systems).

Charging a battery

As already mentioned, there is a vast difference between a scientist talking about a fully charged battery and we ourselves talking about the same. If the SOC (State Of Charge) after a cycle is 99,999% then the battery is not going to loose capacity due to building up lead sulphate. If the SOC is only (!) 99%, then already 10 cycles are enough to reduce the total surface area of the crystals by 65% (see next picture).

In most PV installations the capacity of the batteries will be such that during the night only 25% or so of its capacity will be used (that means there is enough capacity available for additional two days). If the owner is lucky, he has 10 hours of sunshine available for charging the battery and he will be happy when 95% SOC can be reached. It is obvious that the battery will suffer in winter time and the electrodes will be covered thickly with lead sulphate.

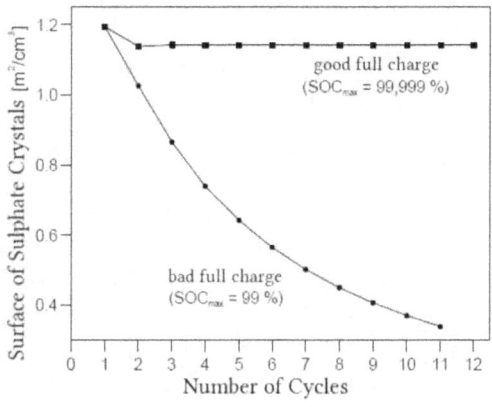

The only remedy is a full charge over a longer period of time. Corrosion in a battery can not be stopped or reversed (although it slows down at low temperatures). sulfating can be stopped and even (over a wide range) reversed, which is why batteries in UPS (Uninterrupted Power Supply) have such a long lifetime. For PV installations this means that somehow one has to make sure that every few weeks the batteries get a full charge, which should last at least 48 hours (or as much time as possible). A practical possibility is to have a battery bank which can be divided plus knowing somebody with a mains socket for the charge. Else you will have to buy a small generator.

From a chemical point of view it is possible to charge a battery nearly without losses. But there are some physical effects which cause losses: resistors and gassing.

The formula for calculating the power losses is

$$P = V * I = I^2 * R$$

The losses caused by the resistors increase with the square of the current where R is the resistance of the battery. If we were able to cut down the current by half the losses would be a quarter. Or in everydays terms: you put two batteries in parallel and you will get [1/4 + 1/4 = 1/2] only half the losses because of the resistance of the battery. If you can't use bigger batteries then you should try to keep the charging voltage only slightly higher than the open contact battery voltage. But on the other hand one does not want to wait endless for the battery to be full or to give the sulphate crystals too much time to grow (to get rid of them would take even more time).

So charging a battery is some kind of compromise!

One often finds the rule of thumb that the charging current (just the number of Amps) should not be higher than 10% of the capacity (just the number in Ah). Thus the regular charging current for a 100 Ah battery should not be higher than 10 Ampere. On the other hand the battery will start gassing heavily when the voltage is higher than 2.4 Volts per battery cell. If it is higher then most of the energy will be used to split up the water molecules and only a fraction charges the battery.

A typical charging cycle starts with a phase with constant current (the charger adjusts this by help of its output voltage). This phase if often referred to as bulk charging and makes it possible to get as much power into the battery as possible in a given time. When the output voltage comes near to the gassing voltage it is kept constant and slightly below. Good chargers now measure the current flowing into the battery. As the battery fills more and more this current will become smaller and smaller. When the value of this current is smaller than 1% of the capacity (with a 100 Ah battery this would be 1 Ampere), the battery is regarded as full.

As we know, scientifically full is not the same as technically full. That's why a last phase follows. Now a voltage of 2.23 Volts per cell is connected to the battery (this is the voltage supplied to batteries in UPS's as well). This fills the last corners (or, if you prefer, dissolves the last sulphate crystals) and delivers the energy used for preventing self-discharge (if an existing electrolyte circulation system is switched on every so now and then then the diffusion of lead sulfate in the direction of conductive materials is supported).

Let's assume we have a PV installation with a 100 Ah battery and overnight 50 Ah are consumed. Some time after sunrise bulk charging starts with 10 Ampere. After roughly 4 hours the constant voltage phase starts which will take 4 to 6 hours. At sundown the battery is charged but it is not really full in scientific terms. In other words, during the summer we will be able to keep the battery in a quite healthy state. But in winter time we can't, not even approximately.

What we need is a battery charger where the user can adjust the bulk current. Another number often found in literature is that a charging current of about 20% of the capacity is tolerable (if really necessary, if for instance you are running a generator for charging the batteries, you can go up to as much as 30%; but read the manual of the battery before you try). What we need to do is replace Old Nick by Lord Harry. The stronger current stresses the battery but sulfating would be worse. We can cut down the bulk phase to two hours and still have a long enough phase with constant voltage. If you bring your car battery to the garage for charging they will normally keep it for 24 hours at least (unless it is an emergency, in which case they can push enough energy into the battery to get you going within an hour or even less).

In other words you should get yourself the best battery charger you can afford (and which makes sense when you look at the price of your batteries). If you connect a charger with a maximum output of 5 Ampere to a 100 Ah battery then the probability that you can say goodbye to it within a short time is relatively high. The charger should allow you the possibility of adjusting the bulk current. It is important to adapt the charger to the batteries you are actually using. Additionally the batteries are losing capacity over time and you might want to nurse your batteries by changing the current from summer to winter use and vice versa. And the charger should know three phases of charging (in technician-speak the charger should be of the IUoU type; first constant current then constant voltage, automatic switching to the voltage for conservation charging). All other options might be nice but are not necessary.

Gassing

In the lead acid battery, water is split up into hydrogen and oxygen - this is something you can not suppress completely. In a normal battery these gases travel through the liquid to the electrodes, build up little bubbles, move up and leave the battery. With the gas the energy needed to split the water into the two component gases will also leave the battery (and that is quite a lot of energy).

Gel and AGM batteries have sealed casings, which is why they should not gas strongly (actually, they can!). Little cracks in the gel and very small tunnels in the absorber mat normally make it possible for the two gases to find each other again and recombine. Now the energy used for splitting the water molecules is set free as heat.

Therefore gassing is normally an unwanted effect since it is associated with a high energy loss. But in two situations the gassing is used on purpose: in order to get rid of stratification (which we have discussed already in detail) and for equalization, which we will discuss in a moment.

Strong gassing starts at 2.4 Volts per cell (for a 12 Volt battery that means 14.4 Volts).

For normal lead acid batteries there is a nice gadget that prevents the loss of water, called recombination stoppers. They replace the normal stoppers and seal the opening gas tight. Inside the stopper is a small catalytic converter; on the surface of the converter hydrogen and oxygen recombine to form water. The effect can be so effective that such a battery can be considered to be maintenance free. The disadvantage is that they seal the opening gas tight, which means that they are not compatible with normal electrolyte circulation system.

Equalization

The process of producing a battery is not that exact that different cells are going to act in an identical way, which leads to the effect that the electrical parameters of the cells drift apart. Let's assume that we have a 12 Volt battery and one of the 6 cells is a bit weaker than the other cells (they are all connected in line). While the other cells have an DoD of just 40% this cell is already at 50%. When the battery is charged once again we come to the point that all cells are full except that particular one. If charging is now stopped (the charger works on the assumption that the battery is full) then this single cell is in good condition to produce big sulfate crystals and lose even more of its capacity.

The problem is that there is no 'natural' mechanism which is going to slow down this process or even to stop it. No, it is accelerated. But how to give an extra charge to a cell when you have no direct connection to it? When the voltage of a fully charged battery is reached the cells are simply not willing to let any current pass through them (if we neglect the self-discharge current). The answer is . . . gassing!

We connect the battery with a voltage which is just too high. The cells protest and start to gas except the one cell which is not full. While the other cells are gassing quietly (Gel- and AGM-batteries) or mightily (normal lead acid battery) the cell that wasn't fully charged gets put on course again.

With a normal lead acid battery there is no difference between gassing to deal with stratification or gassing for equalization; both are done at the same time as one job. Sealed batteries are constructed in a way that they can't have stratification (or only a minimal one) but they are not protected against cells drifting apart.

How can one know that an equalization is necessary? To be honest you can't unless you have a fully fledged battery management system (with a complete monitoring system). If the capacity of a battery goes faster down than statistically expected, then an equalization might be necessary. Since we prefer the pragmatic approach we simply plan to do an equalization every two or three months. If the permitted voltages for each cell are respected (have a look at the data sheet of your battery) then they will survive in full health.

If a really deep discharge occurred or a couple of pretty deep cycles in a row then you should also make an equalization (if you don't do it then one or more cells will age much faster than necessary). If the battery is gassing frequently because of the electrolyte mixing then you don't need to care about equalization.

The voltage during equalization should be 2.55 Volts (others say 2.4 Volts; so have a look at the data sheet of the battery). An equalization should not take longer than half an hour (mild method to be used frequently) or take as long until there are no changes in the output voltage of the battery for two hours.

Efficiency of batteries

In this chapter we are going to look again at our 100 Ah battery. A battery can be seen (in the simplest form) as a resistor (the famous internal resistance) and a capacitor (to store the electric energy) in a simple linear circuit. The internal resistor is the sum of the resistances of the cables, the electrodes and the electrolyte. A 100 Ah battery has

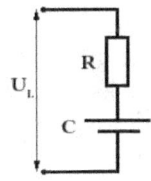

typically an internal resistance of 5 Milliohm (it varies a bit with the exact type of the battery but the exact value is not that important now).

When we connect a charging voltage to the battery then the voltage over the capacitor first rises fast and then the rate of increase becomes rapidly lower and lower. We know that charge carrier with the same polarity push off each other and this force counteracts to the charging voltage which is the reason why the charge slows down. This can be seen in the way the current changes as well. First the current is relatively high after which the current get exponentially smaller and smaller. This explains why we can make a bulk charging at the beginning and use slow charging for filling the last corners.

In this picture the only possibility to generate energy losses is by the current running through the inner resistor. In order to compute these losses we got the formula $P = V * I = I^2 * R$. Generally it is said that for a 100 Ah battery the charging current should not be higher than 10 Ampere; that is $P = 10 * 10 * 0.005 = 0.5$ Watt. We get a bit rougher and use a current of 30 Ampere. $P = 30 * 30 * 0,005 = 4.5$ Watt. That means with such big batteries you will not be able to tell (from the difference of temperature) which one gets charged with 10 Amps and which one with 30 Amps. The difference in losses is simply too small.

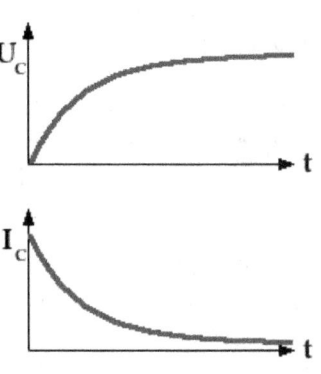

So we have $P = V * I = 12 * 30 = 360$ Watt going into the battery compared with 4.5 Watts losses at this current; when we discharge the battery again with the regular 10 Amps the losses will be 0.5 Watts. So in total the losses are 5 Watts which is 1.38% of the usable energy or an **efficiency of 98.62%**. On the other hand people claim again and again that normal lead acid batteries have an efficiency of maybe 80% (I used this value as well in the book).

The owners of OPzS batteries (the ones which are used within fork trucks) report regularly of efficiencies about 90%. If they were right: where does the rest of the energy goes to? The answer is gassing - the energy is used to split the water to hydrogen and oxygen; later, when the two recombine, this energy will be converted into heat. The consequence is that normal batteries stay cool because the gases leave the battery while sealed batteries get warmer.

The result of this little excursion is that it is absolutely 'legal' to charge the battery in the bulk phase with the highest current the manufacturer allows. The data sheet does not only tell how high the maximum current can be it also tells the inner resistance of the battery so you can easily calculate how much heat is going to be produced during the bulk phase. As long as the charging voltage stays below the gassing voltage of 2.4 Volts per cell (14.4 Volts for a 12 Volt battery) the losses caused by high currents will normally be negligible.

Or to put it the other way round: whether you have losses of 0.14% when charging with 10 Amps or you have losses of 1.38% when charging with 30 Amps is of no real interest! The really important losses you get when you charge a battery with too high a voltage (that is, your battery bank is too small and in order to get the energy into the battery you have to use voltages which cause the battery to gas!) If in doubt: listen to your batteries when they are charging! When you can hear them bubbling strongly you might feel happy because you hear free energy flowing into the batteries. Well, you are utterly wrong! What you hear is burning money. You pay for additional panels and the additional energy is just dissipated from the batteries!

Generally people think that batteries have an efficiency of about 80%. If we make the battery bank bigger and reduce the losses by 10% then we get a funny effect: we spend more money on the batteries and we have to pay the interest rates for them (assumed to 5%). On the other hand we have 5% more electricity output from the batteries. So we save quite a lot of money by spending more (better efficiency and/or smaller installation)!

How to calculate the lifetime of batteries

Standard EN 60254 is meant to make it possible to compare batteries from different manufacturers with various models. These standards define how and under what conditions tests with batteries are to be performed. One of the most important data about a battery is the expected lifetime under certain carefully defined conditions.

Since one normally does not know exactly the conditions under which the battery will have to work later, such a calculation can't be more than just a rough estimate. But the test conditions are standardized and the behavior of different batteries with the same capacity will be quite similar in their general behavior. So at least one will be able to select the battery which will meet ones needs at the best price.

One of the basic assumptions in this standard is that the life time of a lead acid battery is directly related to the amount of energy passing through it. That means that one cycle with 80% DOD will stress the battery as much as two cycles with 40% DOD. Some manufacturers deliver charts on which you can read how many cycles with a given DOD the battery will survive. The diagram below is based on a handbook[1] (page 48) of Hoppecke being one of Germanys biggest battery manufacturers.

1)http://www.hoppecke.de/fileadmin/_hoppecke/content/download/1638/12657/file/Montagehandbuch_geschl_de0113.pdf

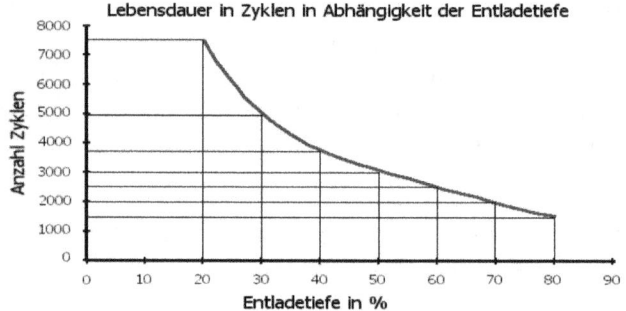

OPzS-batteries are well known for extreme reliability, long life and high resistance against deep cycles. Let's assume once again that we have a 100 Ah battery and for different DOD levels we read the number of cycles and put this information in the form of a table (since we have 100 Ah we can cancel the 100 against the percentage). A DOD of 40% now means 40 Ah. So we get this table:

DoD	number cycles	throughput in Ah
20%	7,500	150,000
30%	5,000	150,000
40%	3,750	150,000
50%	3,000	150,000
60%	2,500	150,000
70%	2,000	140,000
80%	1,500	120,000

We see that the energy throughput of the battery stays roughly the same no matter which DOD we choose to work with (as long as we stay away from DOD's higher than 70%). If there is no diagram coming with the battery there will be at least a statement like: 1,500 cycles with 80% DOD (if it does not actually say 80% DOD then 80% DOD are meant if the description follows the standard). That means that the OPzS battery with 100 Ah will survive a throughput of 150,000 Ah.

But what does that mean to a normal user? The answer is short and simple: any kWh stored in the battery and regained later will cost exactly the same amount of money! Lets assume two neighbors with identical equipment, except for the battery. A has a 100 Ah battery and B's battery has only 50 Ah. A goes down to 30% DOD and B needs to go down to 60% DOD. The rule now says that A's battery will last 6 years and B's only 3 years. Both batteries are of the 12 Volt type, so that one Ah costs one Euro and after 6 years both paid 100 Euros for their batteries. No difference price-wise!

On second sight there is an important difference. With an isolated installation one needs to plan how much time one wants to be able to run on batteries (time of autonomy during bad weather). The capacity of the battery determines whether there is just enough energy for one night or there is sufficient for 4 days of bad weather in winter. Now we come to the interesting point: any kWh from the battery costs exactly the same no matter when it is used.

The only cost difference (if we assume that the additional space for storing the bigger battery doesn't cost anything) is the rate of interest: you pay a higher amount in interest rates for a bigger battery and in turn you get a longer time of autonomy! If an installation is large enough that it can bridge 5 days one will normally work with cycles of 15% DOD; as a result the battery will survive many of these shallow discharge cycles (that means many days of life time). Some deep cycles in between shorten the life in number of days but not in kWh throughput!

If no lifetime longer than 10 years is calculated we come to this roundup: the bigger the battery, the cheaper it will be (related to the capacity). The bigger the battery the longer the time of autonomy without generator use. The bigger the battery the higher are the losses if you do something completely wrong with handling and maintenance.

How much does a kWh from the battery cost?

Up to this point we learned what to do in order to make the life span of the battery as long as possible and that, with this background information, a bit of maintenance and some thought can save us a fair amount of money. Now we come to the point where we ask ourselves "Why we want to have this battery at all"? The answer is that we want electric energy during the hours when the sun is not shining, and the main question is "How much will that cost?" Luckily there is a relatively simple way to answer this question.

We start again with our 100 Ah 12 Volt battery which costs 100 Euros. The data sheet says that it can go through 1,500 cycles. The battery can store 1.2 * 0.8 = 1.440 kWh and do that 1,500 times. Thus it can pass through 2,160 kWh in its lifetime at a cost of 100 Euros and this fact does not change if we change the lifetime in days. That is 4.6 Euro cent per kWh that has been stored and used subsequently. If we include the costs of the battery charger, interest rates and the losses of the charger and the battery we will end up with a cost of less than 10 cents per kWh.

As already said, this price does not depend on whether you want a back-up for two or for ten days. The price per kWh is simply a constant (for the selected type of battery)!

Tip: you have to check the data sheets of the batteries you intend to buy; if the correlation between DoD and number of cycles is not linear, my last statement is not true for your intended installation! With simple traction batteries it should.

Rescue of batteries

The capacity of a battery can best be measured by help of a stress test. Connect the battery for at least 24 hours to a good charger, disconnect and measure the output voltage two hours later. Then you let the battery sit for a couple of hours and you measure the voltage again. If the difference in the voltages is over 0.1 Volt then the battery is ready for the recycling bin.

Now you connect the battery to a load strong enough to really challenge the battery (e.g. connect a 100 Ah 12 Volt battery to a 120 Watt load for the space of an hour). During that hour you measure voltage and current every 5 minutes. The volts you can measure with your multimeter but probably it will not be able to cope with 10 Amps plus 20% safety margin; look for a suitable ammeter or work with a shunt (search in the internet what that is and how to use it). The change in the readings should be steady; else there is something really wrong with the battery.

From these data we can calculate the energy taken from the battery (Volts times Amps times Time). Then we draw a diagram. On the x-axis we put ten hours and on the y-axis we put the capacity. At zero hours we mark the nominal capacity of the battery; connecting this mark with the 10-hour-mark on the x-axis shows how a new battery would function. Then we mark at one hour the nominal capacity minus the just used capacity. Now we draw a line through the two capacity marks and look where the line crosses the x-axis. Number of hours * 10 is the percentage still available (ten hours is 100% and 5 hours is 50%).

So it is no big job to measure the capacity of a battery. If the remaining capacity of the battery is more than 80% of the nominal capacity everything is fine. If it is lower, by definition the battery is near to useless, but that is no reason why we should not continue to use it a couple of years more. When you buy a new battery you return the old one and in principle you get paid the material but not the remaining capacity. So better use it until you think that the remaining time of autonomy becomes too short.

If the remaining capacity is below 50% the question arises whether there is any hope for recovery. If the battery is older than 6-7 years the answer is simply: no! Especially if the battery had a electrolyte circulation system - it simply has passed away.

If the battery is younger and you suspect sulfating you can give it a try. Connect the battery for two weeks to a good charger and let it gas moderately for one hour every day (for mixing the electrolyte). Then you repeat the stress test. If the capacity did not increase drastically the battery is sent to the recycling company. If the increase was at least considerable you might repeat the whole ceremony (as often as you like). But after a while a definite thumb movement - OK or not OK - will be necessary.

There are companies that offer a battery recover service. Since you did the stress test yourself you know the actual state. Agree on a price you are willing to pay for any additional Ah (but remember a new battery costs just one Euro per Ah based on 12 Volts). If it doesn't work out, you did not loose anything. If the company doesn't agree on such terms, better spend your money on a new battery (or a good used one but not without performing a stress test yourself).

That brings me to the topic of old treasures. If you find in the basement an old battery you stored there two years ago, then this battery has snuffed it. If you really haven't got to do anything better you can try to resurrect it. For doing so you will need a charger on which you can select the battery voltage manually (the modern automatic ones would detect a flat 12 Volt battery as a 6 Volt battery). If you can't delimit the charging current to low values (it might be that all the energy is converted into heat and the acid starts to boil) then you should try a small and simple charger which is absolutely short proof. Check every now and then whether the charger gets too hot and try the stress test after a couple of weeks (check regularly the liquid level!).

Everything else, no matter whether patented or not, will increase your losses and this has very simple reasons. All the material that has found its way down into the sump can only be reactivated by help of advanced recycling techniques (probably as new batteries). There is no chemical and no physical way to get these materials out of the sump and distribute it where it would be needed on the electrodes. The same applies to active material which disappeared by corrosion. There is no way to get it back in place!

The electrodes of batteries consist of a highly conductive grid and on this grid a paste was distributed in order to produce porous lead or lead dioxide sponge. Together with the sulfuric acid these are the active materials of the battery. The amount of these materials is adapted to each other and there is only one chemical interaction: from these materials to lead sulphate or back!

If there had been any possibility to recover the battery our long term charging would have done the trick. That means the sulphate is in places (for example in thick crystals) where it can not be reached (e.g. those parts of the electrodes which have no conductive connection any more with the poles). There might be chemical reactions for dissolving the sulphate without doing too much damage to the rest of the battery. Maybe it might be possible even to separate lead and lead dioxide. But there is no way at all to cause these material to form porous layers on the electrode grid. The keyword is porous. To get it back on the grid might even be possible but the big number of square meters of active surface per gram will be down to square centimeters. Sorry, no hope. The capacity is gone forever!

Battery interconnections

Some might remember from their time at school the standard electrode potential of metals. Such a potential exists for all materials which can interact with electrolytes. Lead acid batteries have a potential between 2.1 Volts (fully charged) and 1.96 Volts (nearly flat). Simplifying it all, the potential is 2 Volts. A normal starter battery has 12 Volts because it is composed of 6 cells of 2 Volts each. The cells are interconnected inside the battery so that from the outside one can only see the two poles. Batteries which are not maintenance free have six stoppers one for each cell (after removing the stopper one can refill the cell with distilled water; with maintenance free batteries this is not possible).

The main building blocks of batteries are cells of 2 volt. If a higher voltage is needed then some batteries must be interconnected in line (the minus electrode of the first cell goes out; the minus electrode of the second cell is connected with the plus electrode of the first cell; the minus electrode of the third cell is ......; the plus electrode of the last cell goes out). In steps of two Volts the desired voltage can be obtained.

And there is a second standard way of interconnections and that is in parallel. Here the minus electrodes of all cells are interconnected and all the plus electrodes are interconnected. The output voltage is the same as of any single cell. The parallel connection is used if more energy has to be delivered at a given voltage.

Parallel connection and serial connection can be mixed but you have to make sure that all the units you put in series do have the same capacity (in Ah) and those in parallel have the same nominal output voltage. When connecting two (or more) serial lines in parallel one should connect the poles on the same voltage level with a thin cable; this is not really necessary but prevents the values of the different cells from drifting apart too fast. Concerning technical terms: a cell is just a building block; a battery is a unit composed of cells (or even just one cell) which can not be separated; a battery bank is the interconnection of cells or batteries where cells or batteries can be connected or disconnected.

When connecting a battery bank to a solar installation it will be known which capacity will be necessary. Often the voltage is already given because of the use of certain components. With this information one goes into the catalogs of the battery manufacturers. If they don't have readily available components that are suitable for the installation one has to assemble the required battery bank. That is why the manufacturers don't sell complete battery banks but just the necessary components (big batteries or battery banks that are already interconnected would be too heavy for transport anyhow).

These components (cells) are normally much taller than car batteries, and they come with prefabricated cables for screwed connections; within minutes one can assemble the required battery bank.

This way of doing things has an additional advantage. Sooner or later all technical devices will break. This could be because of wear and tear, or it could be because of material or production faults. If you have a 12 volt battery as in a car and one cell breaks then the whole battery has to be replaced (if the battery costs 100 Euro and one cell breaks then in effect you drop 83 Euros into the bin because you can't separate the bad cell from the good ones).

In a battery bank you can replace any single cell and of course you can check them separately. The simplest way to check is when no load (or only a small load) is connected to the battery. Then you check cell by cell the voltage and these voltages need to be identical (plus or minus a few percent). And you can repeat this measurement when the battery bank is under load. If one cell differs significantly from the others you can already order a new one (or do an equalization charge).

Even if a cell fails completely (e.g. an internal short) you are flexible. You just take the defect cell out of the battery bank. The battery bank was designed to deliver lets say 48 Volts. You take out one cell and 46 Volts are left. No real problem, the system keeps running!

Short summary of facts

There are three main types of lead-acid-batteries:

> *closed batteries, wet cell (with stoppers); FLA (Flooded Lead Acid)*
> In the form of traction batteries they are the ones to be used in PV-installations (for example OPzS or PzS batteries).
>
> *AGM (absorbent glass mat) / VRLA (valve regulated lead acid)*
> Glass mats separate the positive and the negative electrodes and the mat is soaked with acid. If this battery gasses the safety valve lets the gases out if the pressure is too high. Lost water can not be replaced (it could not mix with the acid), so they don't have stoppers.
>
> *Gel batteries*
> The diluted sulfur acid is mixed with silicate and becomes pudding like. It can not flow but ions can move (almost) freely through the gel. These batteries can be used even upside down. In a PV-installation we don't need this option very often which actually makes the batteries much more expensive. Again, these batteries must not be allowed to gas intensely.

Remark: there are AGM-batteries where the electrodes are thin and wound up in coils, and are so thin that sulfur acid can practically not leak through them. These are largely protected from acid stratification. But they are definitely not cheap!

Any kWh which has gone into a PzS battery and out again costs 10 Cents. This may sound peculiar but it is mainly because of the mechanical wear and tear of the battery.

Batteries have three mortal enemies: corrosion, sulfation and high temperatures (high temperatures aid and abet the first two enemies).

Perfect battery temperature: 20 to 23°C

Every 8°C over 25°C shorten the life of the battery by 50%. Batteries used all the time at 33°C will reduce their lifetime by 50%; all the time 41°C to 25%.

Corrosion can not be prevented completely. It can be slowed down by low temperatures and it can be largely prevented by not getting the battery to gas (the oxygen will attack the electrode of pure lead).

Sulfating can be hindered by regularly charging the battery completely. A fully charged battery ages only because of corrosion. When discharging a battery lead sulphate crystals develop. There is no way of stopping bigger crystals growing at the expenses of smaller crystals. The higher the temperature the faster the reorganization. Bigger crystals dissolve disproportional badly.

Any discharge and charge, no matter how shallow, destroys the battery and this damage is not reversible. Deep discharges cause very high amounts of mechanical damage. Just one really deep discharge can be enough to kill the battery!

Countermeasures against sulfating: electrolyte circulation during charging, or at least gassing the battery frequently. Charging the batteries on a regular basis as long as possible (removal of as much lead sulphate as possible). The longer the intervals between charging the bigger the sulphate crystals will be; the longer the last phase of charging takes the better for the battery.

With industrially used traction batteries one can use 70% DOD on a regular base. Cheap grid plate batteries (car starter batteries) should be used only up to 50% DOD since they suffer strongly under mechanical decomposition (the relationship between lifetime and DOD is a bit more complicated than explained earlier).

The numbers in the following table are just for orientation since the information found in the literature differs slightly and depends on the temperature and the used additives for the material (so check the data sheet).

SOC in %	Tension in Volt	Specific gravity in kg/l
100	12.60	1.26
75	12.36	1.21
50	12.10	1.16
25	11.90	1.12
0	11.80	1.05

Charger output for conserving batteries is 2.23 Volts per cell. With a 12 Volt battery that is 13.4 Volt. Above this voltage the gassing gets stronger.

At a voltage of 2,4 Volts per cell strong gassing starts (with a 12 V battery that will be 14,4 Volt). This should be avoided unless it is wanted for acid circulation or a cell equalization (wastage of energy, corrosion and slight wear and tear of the electrodes).

In order to measure the specific gravity of the acid an acid siphon is needed (properly called an areometer). This is just a glass tube with a rubber ball on top as a pump. Within the tube is a floating body with a scale at its side. You read the scale at the top level of the acid and get the specific gravity. You can get acid siphons for a few Euros at any shop selling car spare parts.

Safety at work: acid can cause chemical burns. That is why you should wear protective goggles and rubber gloves when doing battery maintenance, and it is advisable to wear old clothes as well (or a plastic apron). Chemical burns must have water applied immediately. Chemical burns of the eyes must have large quantities of water applied and you need to see the doctor straight away. Sulfuric acid really is aggressive!

If the batteries are not actually standing on the floor you should secure them so that they can't possibly fall down - the connection cables are not sufficient. Traction batteries are not designed to fall down on to a concrete floor from a height of a meter or more without damage; the battery would break and spill its content. So better spend 2 to 3 Euros for a tension belt or something similar.

Often one finds the advice to place batteries in a plastic basin (the acid might leak out). Plastic rubbish bags are much cheaper (spread them, put some pages of a newspaper on the ground in order to protect the plastic, put the battery on top and pull up the side a bit; costs less than 5 Cents per battery instead of 10 Euros or even more).

Metal parts: batteries store such an amount of energy that they can be used directly for electric welding. Therefore all the tools you are using when working at or near the batteries must be insulated (in case of an emergency repair a few layers of electrical tape may do, although this method is not recommended!); if you provoke a short between the electrodes the material might weld with the electrodes and you can't remove it. Then the battery will heat up which might cause the battery to break and spill acid everywhere. Before you start to work remove all jewelry such as necklaces or metallic watch straps; you can never tell in advance how a silly situation might develop.

Filling up electrolyte: the liquid level should be some millimeters above the edge of the electrodes. Transparent casings have marks on the outside. Best is to use distilled water for filling up the battery. If that is not available use demineralized water. In emergencies you can use water from the tap as well. Never use salt water because the battery will start to produce toxic gas!

Ventilation: a gassing battery produces explosive gas. In Germany the DIN VDE 0510 Part 2 regulates how a battery storage room should be designed and how ventilation should be provided. Batteries able to gas must not be placed in habitable rooms (where somebody is possibly going to smoke a cigarette). Hydrogen is very fleeting. In practically all cases with less than 10 kWh capacity it will be absolutely sufficient to have two openings in the wall to the outside (diameter a few centimeters); one near the floor and the other one near the ceiling. If you have any doubts ask the company where you bought your battery; they will definitely know (at least they should).

Another option is to use vent plugs; these are interconnected with a tube which goes directly to the outside. Then the batteries are save to stay wherever you want them.

8. Supplement: battery management

In Germany for quite a lot of people photovoltaics became some kind of hobby and wherever there is a demand there will be a market. Using your PC you can get all the information (even the not so relevant information) from your photovoltaic installation. They can even make remote adjustments while in holidays. And this trend did not stop in front of the batteries.

To me it doesn't make any sense to make statistics about the aging of the solar panels by daily readings. With the batteries this is slightly different because they are quite expensive if you do something really wrong with them over a longer period of time. In a well dimensioned installation it will be necessary to take care of the batteries every four to eight weeks. That is checking the level of the electrolyte and refill if necessary, to check the voltage of the different batteries (cells) and decide whether an equalization charge might be necessary. With an installation for a one family household you might spend 15 minutes per month with this job (probably you will spend much more time getting your car cleaned).

Until now there is no reasonable system on the market which could do the refilling of the batteries with distilled water reliably and automatically. So this job is for you (unless you spend an unreasonable amount of money for maintenance free batteries). There are many owners of PV-installations who tried out automatic filling and there are lots of reports why these systems found their way into the bin (mostly after the user had to mop up some problem).

One part of the battery management is useful and can be done automatically (the bad news is that I did not find a product doing the job in the way necessary; so this chapter is about a crafts project). For the explanation I assume that we have a battery bank consisting of cells with 2 Volts which are interconnected in line. From the supplement about batteries you will know that these cells can't be manufactured in a way that they are identical. When they work, that is being discharged and charged again, then the differences between the cells increase slowly. In a complete battery bank you would react on this (after a while) with an equalization charge and force all cells to (nearly) equal values.

Lets assume you have a battery bank with 48 Volts which consists of 24 cells. 23 are fully charged but one is not. Now we have to connect the battery bank with a voltage of about 57.6 Volts. All cells start to gas but one; this one would be charged! When this cell starts to gas as well then the equalization charge can be stopped; all cells (should) have the same actual capacity.

But what a wastage of energy! Only 4.2% of the energy is doing something useful; the rest is converted into heat or into oxyhydrogen gas which goes into the surroundings.

For making a cell-wise equalization by hand you will need a voltmeter and a battery charger which is able to charge single cells. In order to prevent any kind of electrical problems the output of the charger must have no galvanic connection to its power supply (there must be a transformer in it so that no electron can go from the input to the output or vice versa). Then it is possible to charge a single cell without any interference with the rest of the battery bank (the current of the charger is superposed to the current running actually through the cell; if the battery bank is actually discharged then this cell is discharged less fast; if the battery bank is actually being charged then the cell is charge slightly faster).

First one would measure the voltage of the battery bank which could be 50 Volts. If all cells were equally full then they all would have a voltage of 2.08 Volts. If we allow a tolerance of 3% then the voltage of the cells must not be lower than 2.02 Volt or higher than 2.14.

Then we measure the individual voltages of the cells. If the voltage is lower than 2.02 Volts then we need to charge it. So lets say we charge the cell for 5 minutes, then we switch off the charger and we measure the voltage again. This game we repeat until the voltage of the cell is within the limits (with a little bit of training you will be able to replace the 5-minute periods with the correct time; another idea is to set up a table with the differences in tension and the corresponding capacity; then you can calculate the necessary time directly). Was the voltage higher than wanted the cell could be discharged by help of a resistor.

After a while you will know which cells are misbehaving and prevent them from aging accelerated. But my idea is to do it fully automatic which will save you all the time (after spending a lot of time building the 'equalizer').

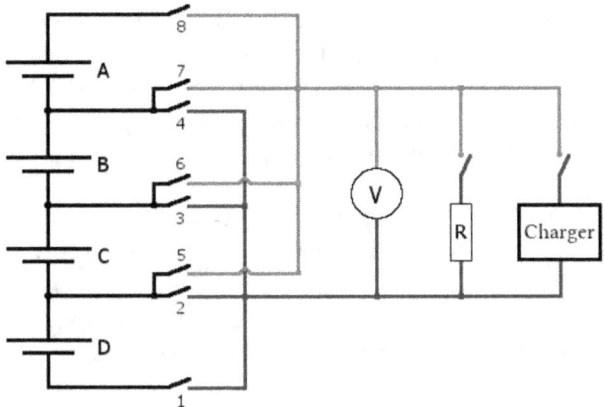

In the equalizer we have an array of relays for connecting the charger with the different cells or a resistor and we need to have some kind of micro processor (Arduino or similar). In order to measure the total voltage of the battery bank the contacts 1 and 8 have to be closed. Now the voltage can be computed which each of the cells should have. In order to connect with cell D the contacts 1 and 5 have to be closed (and all others need to be open!). If the voltage is too high then the cell can be discharged and if it is too low it can be charged. If the voltage is in the correct range the next cell is checked.

Charging and discharging influences the voltage of the cell. Therefore it is necessary to stop in between, recheck the total voltage (the current through the battery bank might have changed and that would cause wrong results) and continue. How long the equalization takes depends on the size of the cells and the power output of the charger (and of cause on how far the cell had drifted away).

With such a device there would be no problem to make an equalization on a daily bases and with an optimum of energy the job is done already before a cell can age faster than the others. In combination with a electrolyte circulation system there would hardly be anything more one could do for the health of the batteries.

9. Supplement: electric fences

On first sight electric fences and a book about photovoltaics don't fit together. At second sight they fit very well. The reason is that normally grazing land and the next socket are far apart. Therefore one needs electric fence controllers which are battery driven and on a regular base one has to exchange the batteries (normally you got one battery at the grazing ground and another one is in the garage to get charged).

Normally (at least in Germany) you have to control the fence on a daily basis but schlepping heavy batteries around is no favored pastime. Therefore the question pops up whether it could be possible to charge the battery by help of a solar panel. In order to know how big such a panel will have to be one needs to know the consumption of such an electric fence controller.

While searching for such data one can find out that most manufacturers don't want to intimidate their (to be) customers with technical details. So again we have to start with the basics. Hopefully you remember that DC voltages higher than 120 Volts and AC voltages higher than 50 Volts can be deadly. In the advertisements we then read that these devices produce between 2,000 and 10,000 Volts. If that is true everything touching the wire must be grilled. Naturally not; the authorities responsible for technical inspection make sure that these devices are not dangerous (quite a lot of snails can't talk about different experiences any more).

So, lets have a look at the working principle of these fences. All around the grazing ground posts are set (or existing ones are used); to each of these posts isolators are attached and then a wire or a conductive ribbon is spanned (the animals can see the ribbon much better). Hopefully you remember that an electric circuit has to be closed so that something can happen; but here we only have a conductive ribbon floating about a meter above the ground. That can't do the trick.

The electric fence controller has got two connectors; to one connector the ribbon (wire) is connected and to the other connector we join an earth anchor (mostly a simple iron peg which is hammered into the ground). Since the ground is conductive (water with different salts is an electrolyte and therefore conductive) we have under the electrically free floating ribbon a large conductive area. Now comes the first part of the trick: normally there will be no current between wire and ground. The fence controller produces an electric pulse which travels over the wire and looses energy while doing so. The few thousand Volts are the open contact voltage!

If now a cattle (or any other animal) comes too near to the wire then it closes the electric circuit and a current flows from the controller over the wire, over the cattle into the ground and via the earth anchor back to the controller. When the farmer comes wearing his rubber boots and by chance touches the wire nothing is going to happen. That is, because rubber is an excellent isolator and no current is flowing. If he does the same on Sunday (now he wears his good shoes with the leather soles) then he will get an electric shock and possibly decide that Sunday is not for work anyhow.

Now we come to the question why this electric shock is absolutely unpleasant (the cattle shall have a huge respect from the wire after touching it just once) but not deadly.

In order to reduce the effect to just being unpleasant two mechanisms are utilized. Within the controller a capacitor is charged and when it is full then this stored energy is used to produce the high voltage pulse. The amount of energy that could possibly be transferred onto the cattle (or farmer) is delimited by the size of this capacitor. Additionally such a cattle has got an electric resistance of about 500 Ohm. The voltage breaks down at the moment of contact.

DIN EN60335-2-76 says that the time period between two pulses must be at least one second and with a standard load (DIN cattle with 500 Ohm?) the energy of the impulse must not be higher than 5 Joule (in order to make really sure that the installation works at all the minimum impulse energy is normally set to 0.5 Joule; if the amount of energy transmitted by help of one pulse is much higher than 5 Joule then it might happen that the manufacturer has an intense conversation with a judge).

One Joule is one Watt second. With a pulse energy of one Joule per second in total 86,400 Joule are given onto the wire per day (normally the fence has to be active during 24 hours). That equals 24 Watt-hours or 0,024 kWh per day. So I had a look at the data sheets of quite a lot of electric fence controllers and compared the electric consumption (some companies don't make a secret out of it) with the pulse power. The devices I found had an efficiency of round about 50%. As a basis we can calculate with 0.05 kWh per Joule impulse power per day.

If in a given case there are no reliable data available then you have to measure. Measure the battery voltage one day and repeat this a few days later (when you can see a clear difference of the values). In the chapter about batteries there is a small table giving voltages and rest capacities. From the difference of the voltages you can (exact enough) calculate back to the capacity taken from the battery during this period and with that we have the consumption per day.

Let say you select a 50 Ah battery (this type of battery most people can still carry without breaking their back) to power the controller. This battery (50 Ah * 12 Volt = 0,6 kWh) could power the electric fence during 12 days without a recharge (I neglect that there are winter days as well when the battery looses quite a lot of energy and that one should not discharge a battery that far).

In December one can calculate in northern Germany (data of Hamburg from PVWatts) with a radiation of 0.63 kWh per day per square meter (inclination 53°). We want a panel with 100 Watt (peak) which costs roughly 100 Euros (transport included) and got an efficiency of 17%. I just selected one with the size of 117cm x 55 cm = 0,64 square meters. It delivers 0,63 kWh * 0,64 m^2 * 0,17 efficiency = 0,068 kWh per day in December. The output voltage is rated with 18 Volts and we use a simple PWM regulator with an efficiency of 66% and get during a winter day 0,044 kWh to the battery. Now we take the efficiency of the battery with 80% into account so we have usable 0.036 kWh per day left. Since we need 0.05 kWh per day it is clear that we will have difficulties during winter time.

We come to the conclusion that our installation would work perfectly if it were not for December and January. Now we have three options: a) we spend more money and get us an MPPT controller; b) we take a bigger panel and a bigger battery or c) we try a kludge (nothing lasts longer than an interim solution!)

The solutions a) and b) are no challenge to you any more so I explain about solution c. Here we make use of the fact that we have to inspect the fence on a daily basis anyhow. We buy a little 12 Volt battery with 10 Ah (you get them for electric lawn mowers for about 25 to 30 Euros). A battery charger you need to have anyhow for the big battery so we don't have to buy a new one.

The 10 Ah battery can store 0.12 kWh. So we could take the full small battery to the controller and replace the big one. This we take with us and charge it for 24 hours and swap the batteries the next day again (if really necessary the small battery could be good for nearly two days). Afterwards you have at least a week until you have to repeat this trick.

If you don't like to carry around the big battery then there is a brutal option. You take the full small battery with you and just connect it in parallel to the big battery (take thicker cable since there will be relatively high currents). If the small battery is really full and the big battery is down to 50% (that is about 12 Volts) then nearly half the energy of the small battery will be transmitted to the big one; that is enough energy for a complete day without sunshine.

Such a 10 Ah battery weighs between 3 and 4 kilos so you can carry that one around without any difficulties. The time necessary at the fence controller is negligible because you only have to connect two prepared crocodile clamps.

The solar panel will last 20 years and the solar controller will as well. The batteries will probably reach old age as well because - except for two months in winter time - they are preserved in perfect state for the rest of the year. Even cheap batteries should last 8 to 10 years.

Last problem is now: how to know that you have to use the additional battery and when to stop that trick again? I think the decision is between 12 and 12.2 Volts (so you will need a voltmeter). If the voltage of the big battery drops below that value then you start with the additional charging. If the value is higher again you can stop (voltmeters with good enough displays you can get per mail order from Hong Kong for 6 to 7 Euros postage included; so you can just leave it on top of the big battery; in case somebody steals it you just buy another one).

10. Supplement: inverter for Minigrids

Actually all inverters are build either for islands (generally small islands with just one inverter) or they are for direct feeding into mains (in Germany nearly all of them). This chapter deals with the question of how inverters should look like (and function) suitable to deliver electric energy to those who don't have a connection to mains yet or where mains is completely unreliable or much too expensive. These inverters would not be meant to support the grid or to use it at all (normally).

Ultimately this chapter is written in order to arouse the attention of the manufacturers; but it could be interesting for other readers as well. If, while reading, you find out that you do not understand anything except the single words then this is no reason for being upset; I want to inspire those technicians who do know how to design an inverter and how to build one.

In the first place it is necessary to explain (simplified) how the grid works. Then I'm going to explain how a Minigrid could function (the integration of some to many inverters but without the use of classic generators).

Basics: the grid

The backbone of the grid are the synchronous generators (before the Energiewende started practically all the electric energy in the grid was coming from them). Synchronous generators deliver a three phase voltage. At the (big) electric consumers four cables are coming out, the three phases and the neutral line. The tension between neutral and each of the phases is 230 Volts (in big parts of Europe). The tension between the phases is 400 Volts. The frequency is 50 Hertz (over shorter periods of time there might be some deviations but in the long run the frequency is as stable as an atomic clock). Between the signals on the three phases there is a shift of 120°.

As already described in the chapter about electrical machines all these generators are coupled via the cables as if there were a mechanical connection and such a grid has got a gigantic power. If you want to connect an additional synchronous machine with the grid you first have to accelerate it until it rotates with the correct speed. Only when speed and phase are correct it can be connected to the grid (if either rpms or the phase displacement were not correct then there will be a pronounced bang and you will have to spent quite a few Euros for new fuses).

In such an interconnected grid the amount of generated electric energy must exactly equal the amount of consumed energy (else in some places energy would simply disappear and in other places, out of thin air, electric energy would reappear and that is not possible). You could visualize this by thinking that the generators press on the sine signal from the left (deforming it) and all the consumers try to press from the top and the right side (deforming the signal as well). If the totality of the generators got more power than the consumers then the generators will start to rotate faster. When they rotate faster their output voltage will be higher; the equation $P = V^2 / R$ gives the consumption of the so-called ohmic consumers and it goes up rapidly with the voltage. Because of the higher frequency the motors will rotate faster and normally they will consume more electric energy (if a pump rotates faster the consumption will increase more or less linear with the rpms). That is, the resistance of the consumers increases and we have got a balance again. The electricity delivering companies have very limited influence on who is going to use how much power so the regulation has to be done mostly on the generator side.

If deviations of frequency and voltage are acceptable the regulation of such a grid is not too complicated (unless there are unforeseen situations). The interesting question is how big deviations are allowed to be. If I feed an normal PC with 230 Volts but with a frequency of 0.5 Hertz then it will for sure not work at all; if I offer the same voltage with 55 Hertz it will work without any difficulty.

IEC 60038 defines that the voltage might differ by +/- 10%. In respect to the 230 Volts we normally work with that means that all voltages between 207 and 253 Volts are allowed. That means that all the appliances sold within the EU have to be able to deal with all voltages within this range (without complaining).

DIN EN 50160 defines that all frequencies between 49.5 Hz and 50.5 Hz are absolutely within range. If the frequency goes below 49 Hz then (in a predefined sequence) electric consumers are disconnected. If the frequency goes below 47.5 Hz then, for self protection, the power plants are shut down (emergency shut down). Then we might get a nice (Europe wide) black out. So the frequency must not differ by more than 5% (up or down). All appliances have to be able to deal with these deviations as well.

Now we need to have a closer look at the way how different devices deform the electrical signal on the line (this explanation is not absolutely correct but gives a good understanding). Grossly we have within a grid synchronous generators, synchronous and asynchronous machines and normal ohmic consumers.

I already explained that the generators push from the left and try to increase the frequency of the sine. All generators try to cave in the signal from the left. The motors try to resist the acceleration and cave in the sine from the right. The ohmic consumers don't care about the frequency; they just try to flatten the sine. What is important for the moment is the knowledge that everything connected to the grid tries to (de)form the sine.

Devices in a Minigrid

Practically all devices within a 230 Volt grid are mass articles. As we have already seen in the chapter about fridges it is very expensive to use appliances needing different voltages. So any Minigrid has to adopt to the local market (mostly used voltage). But there is one big difference: a Minigrid does not need to be compatible with a normal grid! As long the power can be delivered within the boundaries just illustrated (within the EU: frequency +/- 5% and the voltage +/- 10%) than everything is fine.

Then one does not need any special makes and the inverters could be sold from the shelf as well so they might be cheaper as well.

What exactly is a Minigrid

An inverter feeding into mains must work synchronous with the grid (if there is no external voltage clearly detectable then the inverter must switch off immediately for safety reasons). So if you connect an inverter meant for an island then you will have to dispose quite a lot of (formerly expensive) scrap. The reason is that the inverter tries to dominate a few

Terawatts with a few hundred Watts and it can't win. If you connect two inverters meant for an island then, with a bit of luck, only one of them is going to die. The reason is that they have no possibility to synchronize so they work full power against each other.

A Minigrid is established by interconnecting two or more inverters which are synchronized and therefore they can work within the same grid. There are a couple of possibilities in order to realize such a synchronization and I propose that there is an additional interconnection between the inverters in order to establish this synchronization. So we have the cables for the 230 Volts and we have additionally an AC voltage (lower than 50 Volts so that even unskilled workers are allowed to lay these cables) which is the leading signal for the 230 Volt output.

The higher the voltage of this signal is the higher is the protection against disturbances. When transmitting this signal over a two-strand cable possible disturbances are induced on both cables with the same height. When working with the differential voltage between the two strands then the signal can be transported over longer distances (some hundred meters) without significant distortion. It would make sense to use a voltage of 23 Volts (division ratio of ten). In that case the 'demand' voltage might change between 25.3 Volts and 20.7 Volts.

There is no reason why a Minigrid should not have three phases; so we need an electronic circuit which delivers three different voltages, shifted by 120° (if only one phase is needed the other two signals are simply ignored). This voltage is connected through to all the inverters which power this phase. Depending on the power output of the amplifier there could be dozens of connected inverters or even hundreds or more.

The inverters interconnected in this way all the time compare the voltage on the power line with the voltage on the signal line (taking the 1 : 10 ratio into account) and try with the locally stored energy to support the voltage on the 230 Volt line. When the energy reserves are used up they can't support the waveform any more and the signal on the 230 Volt line gets dented.

Since we are not using electrical machines as generators within the Minigrid the left flank of the signal will not be caved in. It will be absolutely normal that there are electrical machines connected, so they are going to cave in the right flank of the signal. For sure there will be ohmic consumers and they will try to flatten the signal. What is important for us is the fact that motors and ohmic consumers will change the effective voltage of the output signal within the Minigrid.

Regulating a Minigrid

At the compound grid we saw that the voltage and the frequency are allowed to vary a bit and the main control variable is the frequency. Since it is by far more complicated to control a system using two independent parameters the Minigrid is controlled by help of the voltage and the frequency stays fixed (it is pretty easy to measure a frequency but it is far easier to measure and compare voltages).

First we have a look at pure ohmic consumers.

If the appliances want to consume more energy than the totality of the inverters can deliver then the voltage must go down and following $P = V^2 / R$ less energy has to be delivered. Naturally the voltages must be kept within the +/- 10% range.

If the appliances demand less energy then the totality of all inverters within the Minigrid can deliver then the voltage needs to rise and following the formula $P = V^2 / R$ more energy is consumed. Naturally the voltages must be kept within the +/- 10% range.

In a perfectly designed Minigrid the voltage would be 230 Volts at normal irradiation. Demand and supply would be in balance and there would be about 10% reserve. How the regulation functions in more detail will be explained in a moment.

In a compound grid the generator side is regulated and only in cases of emergencies they start to jettison electric loads (normally big consuming devices). Within a Minigrid the possibilities to generate more electric energy are extremely limited. Therefore the only solution is that the disconnection of consuming devices is something absolutely normal.

In a compound grid there are technicians who work 24 hours per day at the problem to generate a perfect match between demand and supply. This is work which can't be done for a Minigrid. Lets say such a Minigrid generates 100 kWh per day and each costs 10 cents. Then the installation earns 10 Euros a day and approximately 9 Euros are for payback and interest rates. Even in extremely poor countries you could not find somebody who is willing to work all day for one Euro in a qualified position. So we need a different solution.

The solution is to include the jettison of electric loads already during the planing phase and to make it idiomatically.

Lets start with the standard situation in not (completely) developed countries. When the sun is shining there is electricity and if it is not shining then there is no electricity (unless some energy is stored in batteries). Over the day the supply might change drastically. Since a solar installation costs every day the same amount no matter how much electricity was generated and used, such a Minigrid should be designed in a way that all the time some (not really necessary) consuming devices are in standby. What we need are priorities!

Already during the planing phase a list is set up giving the importance of the different consuming devices. In the morning, shortly after sunrise, the most important consumers are switched on and while the sun moves up in the sky (and more electric energy can be delivered) the 'less important' devices are switched on. If a cloud passes by the less important devices are switched off again.

What at first sight looks like a tremendous effort for collecting all the information has got an absolutely simple solution. Each and any circuit which could be connected with the Minigrid has got something I simply call a 'switch-box'. In this box we have a simple and adjustable voltage comparator. If the voltage on the Minigrid is lower than the reference voltage then the circuit is not connected to the Minigrid and the appliances have no electricity. If the voltage rises above the reference voltage then the circuit is connected with the Minigrid and the appliances get juice (such a box might cost about 2 to 5 Euros). The voltage comparator has a built in hysteresis so that the switching is stable (without a hysteresis we would have the situation that an appliance is switched on; this causes a drop of voltage which in turn is the cause of switching it off again).

Strictly speaking we got with a Minigrid an absolutely primitive version of a Smart Grid. In a smart grid the client is requested to have a look at the actual price and to decide when to make use of the energy supply. In a Minigrid the appliances are only switched on when there is surplus energy. How to know when this is the case? We can tell from the height of the voltage!

First a simple example for the use of such a Minigrid within the 37° belt, a sewing factory. Shortly after sunrise the minimal lighting is switched on (the workers have enough orientation for doing the job preparation). Then row after row the sewing machines are switched on including the working light. Then the offices get connected and when everything is running the surplus energy goes into the air conditioning. During the afternoon the appliances get switched off in reverse order.

In exchange for cheap electric energy (5 cent or even less per kWh without any kind of subsidies) one will have to develop a bit more flexibility. In most countries with such a lot of sunshine this flexibility can be found anyhow.

The voltage in a Minigrid

In compound grids the frequency and the voltage are used in order to regulate the balance between supply and demand (mostly without forced jettisons). In our Minigrid we don't have any synchronous generators (that would make things unnecessarily complicated) so we can keep the frequency constant. If a micro controller is used in order to generate the model waveform for the Minigrid then the frequency will vary possibly in the per mill area; so the load connected to the Minigrid has no influence whatsoever on the frequency so synchronous motors in the Minigrid will rotate steadily.

The very basis of the regulation is that each inverter tries to support the signal on the line as good as possible (following the model wave form on the extra line) as long as it got some energy left. If we have a balance between supply and demand then we will see a technically perfect sine on the line.

Now we come back to the building block which generates the three sines. This component has to get a second job: it has to calculate the difference between the model wave form and the wave form on the line. If this difference is negative (the voltage on the line is lower than the model voltage) then there is not enough energy flowing from the panels into the inverters and they don't have enough power to support the actual voltage in the requested way. So the voltage of the model waveform has to be lowered.

This causes the voltage on the line to go down as well and the ohmic consumers will lower their consumption. The voltage of the model signal is lowered until the form of the signal on the line is similar enough (the summed up difference is small enough). On the way down the voltage gets support from the switch boxes because they start to disconnect one consuming device after the other. During the installation of these boxes a reference voltage was set (can be changed easily later) and when the voltage on the line drops below this reference then the appliances are switched off which in turn supports the signal waveform on the line (advanced switch boxes might give a warning some seconds before actually switching off).

When the voltage drops below 207 Volts then all appliances are switched off because voltages below this value are not 'legal' and a correct function could not be guaranteed (since motors dent in the wave form from the right hand side they reduce the area of the wave form as well; so the effect of motors is accounted for as well).

If the difference of the area of the two wave forms is very small then the voltage of the model wave form is increased. This increases the consumption of all the ohmic consumers in the Minigrid. The maximum voltage in a Minigrid is 257 Volts AC.

We have a linear correlation between the difference of supply and demand on one hand and the voltage on the line. And this makes the regulation of a Minigrid a fairly simple job. The tripping voltages of the switch boxes can be selected within the full range between 207 and 253 Volts. The lower the reference voltage the more important is the appliance!

There is one serious problem during the planing of such a Minigrid and that are appliances with very high consumption because one should make sure that there is enough reserve capacity so that the Minigrid continues to works.

Switch box and switch-over box

Each box consists of three components: a small power supply, a voltage comparator with hysteresis and a relay. The little power supply is directly connected to the Minigrid and has a stand by consumption of just a few Milliwatts.

The relay in the normal switch box is of the 'no' type (normally open); if the voltage on the Minigrid is higher than the reference voltage then the circuit is connected to the Minigrid. If the voltage drops below this level minus hysteresis then the circuit is disconnected again.

The switch-over box contains a changeover relay instead of the simple closer. An appliance could be connected during the day with the Minigrid and when the voltage in the Minigrid drops down too far the appliance is disconnected from the Minigrid and connected to mains (we were discussing something similar with the air conditioners).

These simple boxes could be price-wise in the area of 2 to 5 Euros.

It is technically no problem to make sure that the sine within the Minigrid is synchronous with the sine of a mains cable. If that is done, even inductive loads (like motors, fridges and freezers) could be switched over from the Minigrid to mains or back without causing voltage spikes. If out of whatsoever reason it is not possible that the two voltages are synchronized then the switch over must be done in two steps (first disconnect from one grid and when this is definitively done connect to the other grid).

Requirements for inverters within a Minigrid

Actually I do not know of any inverter fulfilling all listed requirements (so they are maximum requests). I think if a company starts to build inverters fulfilling all these requirements (and realizing Minigrids as described before) they would have very good chances to develop a huge market (unless they are asking prices which are not justified by the product).

- There must be a galvanic separation between the AC and the DC side so that the safety at work on the DC side can be guaranteed.

- The DC input voltage is lower than 120 Volts so that unskilled workers are allowed to install the panels, the cables and even the DC side of the inverters (not only legal safety but technical safety guaranteed)

- When starting with safe 100 Volts and pleasant 20 Ampere on the DC side then a simple string inverter should be able to deliver 2 kW.

- It must be possible to scale up or down running installations (without switching them off)

- For the developing countries inverters of the following sizes would make sense: 100 Watt, 200 Watt, 500 Watt, 1 kW and 2 kW. Since Minigrid string inverters can be interconnected without any problems in parallel all demands up into the Megawatt range could be satisfied

- Inverters for Mingrids must be able to deliver for about 300 Milliseconds a power ten times as high as the nominal power (starting current of electrical machines); this could be realized either by help of small batteries or (super-) ECAPs

- The inverters need to be mechanically and electrically as robust as could be

- The internal design of the inverter needs to be as modular as anyhow possible so that repairing on the basis of module exchange (cannibalism) is possible

- Inverters for the 37° belt need to be dust and spray water tight. Possibly cooling concepts with water cooling or the use of heat pipes plus fans might be necessary

- In the 37° belt temperatures of 40 to 50 °C are not seldom; so a working temperature of 60°C or more is necessary

- The installation, maintenance and repairing must be that easy that a (maximum) 5 days course is sufficient to learn everything.

Carrier Sense Multiple Access / Collision Detection

It is already some decades ago that the American military service started to use automated meteorological stations (so that was before weather satellites were launched). Each weather station had a radio transmitter and a time assigned when to transmit the data. The system worked relatively well but there was a big problem: the clocks at that time were not accurate enough. Normally it took only some weeks and two stations started to transmit the data at the same time and it was not possible to receive the data from any of the two (please remember that at that time computers did exist but they were very expensive and energy hungry).

Then somebody had the brilliant idea to modify the transmitter in a way that it only started to transmit data when no other transmitter was on the air. In that case the transmitter would just wait a few moments and then start its transmission. This method (called Carrier Sense Multiple Access) brought big advances but it still happened that two transmitters interfered with each other.

It was necessary to improve the method a bit. First the station listened whether another station was already on the air. In this case it just waited a certain time (given by chance) and tried again. If no other transmission was in progress then the station started with its report; the transmitted signal was compared with the received signal; if there was no disturbance there was no other station transmitting as well; if there were disturbances both stations stopped the transmission and the next try was delayed randomly a short time.

Later in history this method was again used when computers were interconnected via the Ethernet (10 megabit per second were unbelievable fast at that time). So I think one can say that this method proved during the last 50 years that it works reliably if the system utilization is not much too high.

Now you will have the absolutely justified question what this might have to do with photovoltaics. Strictly speaking it has to do with the system utilization. Lets assume you are running a refrigerated store house within the 37° belt. The overall system design was made in a way that during the day enough cold is generated that it will last until the next morning (plus some security margin; you will use phase change cold storage). If you have too much money you will use batteries.

Now we have the following two problems: when a compressor starts it will need ten times as much energy than the nominal wattage and in all the cooling chambers it is slightly too warm and all the compressors want to start the moment they get electricity in the morning. Normally the compressors have a duty cycle of about 10% and the system design was made to satisfy this consumption (plus some reserve). But now in the morning we have a complete overload (all compressors want to start and they need for a short time ten times as much energy than normally). The result is a complete system break down.

What is going to happen if we upgrade our switch boxes with CSMA/CD? Whenever the voltage, out of whatsoever reason, is too low, no consuming device is trying to get more energy. When the voltage rises the quarreling starts and one compressor will win by chance. All other compressors are caught in a waiting loop. Then we might have the situation that the voltage is just high enough to justify the compressor to start and just by chance a second compressor starts as well. We have a collision because the voltage drops below the wanted level and both compressors stop to work. After a random time both will try to start again.

Scientific research concerning the Ethernet showed that up to a system utilization of 70 to 80% a nearly perfect data transmission was guaranteed. If all the compressors get an individual priority (additionally depending on the temperature in the refrigerating chamber) then with an absolute minimum of effort (no central supervision and controlling) we get a near to perfect energy distribution.

So even if it gets a bit more demanding the Minigrids are a perfect solution for all parts of the world where there is no cheap and rock solid energy supply (you would pay about 25 to 50% of the price which has to be paid in developed countries; nice advantage. Use it!).

The fantastic thing is that it is not necessary to start a single subsidy program. The necessary components are already cheap enough. What will be necessary are training programs but they are by far cheaper than subsidies (and they last longer).

Minigrids with three phases

A single phase Minigrid is much easier to administrate than a Minigrid with three phases. So single phase Minigrids will be preferred. When the currents start to be 'unhandy' then one will start to split the Minigrid into different Minigrids. If you have to supply energy to three phase electrical machines then you have no other choice: you have to realize a three phase Minigrid.

The little electronic circuit generating the model wave form does not simply generate one sine but three which are shifted by 120° to each other. Within the compound three phase grid the aim is to distribute the load as even as possible to the three phases (an imbalance too big might cause damages to the generators). In a three phase Minigrid the situation is a bit different since the power going over the different phases can be completely different without causing any problems.

In a three phase Minigrid we need to have switch boxes as well. If a three phase machine can work without any problem when the voltages on the different phases differ by +/- 10% then there is no problem (the machine is switched off if one of the three voltages drops below the reference value). If the electric machine can't cope with such differences then the regulation will be much more complicated (a more robust machine might be cheaper).

11. Supplement: some astronomy

Initially I wanted to write a little (?) program so that everybody should be able to calculate the basic data for a photovoltaic installation. This program became the victim of my mental red pencil (I wanted to get the book ready and out) and additionally it turned out these calculations are not of prominent importance any more (the panels are too cheap and the input data might deviate more that 20% from the true data). Anyhow, it would make me happy if somebody picks up that idea and turns it into a public domain program (contact me via my homepage). For all others this chapter is more of academic value and you don't really need to read it (it might even be repelling).

The orbit of the Earth around the Sun

If the Earth were the only planet in the system then it would run on an elliptic path around the sun. Since the influence of the other planets is very small (I doubt that it is possible to detect their influence in the strength of the solar irradiation on Earth) we can neglect it. The same is true for the influence of other stars or galaxies.

So I just check the influence of the Moon. Earth and Moon have a common gravitational center and it is this center which moves around the Sun on an elliptic path. One can measure and calculate that this common gravitational center lays within the Earth about 3,000 kilometers away from the center of the Earth but still 1,700 kilometers below the surface. So when the Moon is on the line to the Sun then the Earth is 3,000 kilometers nearer to the Sun. The distance between Earth and Sun is about 149 million kilometers; the back and forth movement because of the Moon can be neglected as well.

So the Earth moves on a perfect elliptic path. If we had the special case of a circle then the distance would be all the time constant but we don't have that one. When the distance is smallest than we have summer time on the southern hemisphere (we come back to that in a moment) and the Earth is flying a bit faster. When we have summer time on the northern hemisphere then the distance is bigger and the Earth flies slower. This has the consequence that the length of the solar day is not exactly 24 hours; and this deviation as well is too small to have any interesting influence on a solar installation.

Another consequence of the elliptic path is that the intensity of the solar irradiation on the Earth is more intense when the distance is smaller (when the distance is shortest then one can receive outside the atmosphere 1,412 Watt and with the longest distance it will be 1,322 Watt per square meter being in a right angle to the Sun). In the years average we receive 1,367 Watt and we have a deviation of 3.3%.

This deviation could be ignored but there is a relatively simple way of taking this influence into account. We only have to multiply the 1,367 with -45 * sin(number of the day). The 45 we get from 1,367 - 1,322 and what is meant with number of the day we will see a bit later in the chapter "A bit more Astronomy". The effect of this term is that for our calculation the path of the Earth is now a perfect circle.

The axis of this rotation is pretty near to the center of the Sun. Additionally the Earth circulates around another axis which goes through its own center (if we neglect the moon). The result of this rotation is that we have day and night. The interesting thing now is that these two axis are not in parallel. Between these two axis is an angle which can be seen, at least for our purposes, as constant over time. This angle is called the ecliptic.

The value of this angle is about 23.5° and this angle is the reason for the seasons of the year. One could say that the rotational axis of the Earth is adjusted to the center of the universe and that it stays in this orientation while it circles around the Sun. When it is summer on the northern hemisphere then all the time it is tilted 23.5° in direction of the Sun. That means on the other hand that the southern hemisphere is tilted away from the Sun all the time and that they got winter. Half a year later the situation is reversed.

So for any place on the world (if it is situated between the two polar circles): when you measure every day the angle of the highest point of the Suns path over the horizon then the difference between the biggest and the smallest angle will be 47° (two times 23.5°). This swing range is by far too big to be ignored. How we include this factor in our calculation follows in a moment.

The time of day

There are some regulations around the world which have an influence on the time we read from our wristwatch (for example the distinction between summer- and winter-time). Since we are not going to do some navigation we simply ignore all these regulation and define that 12 o'clock noon is at the moment when the Sun is at its highest point. For estimating our yield of solar energy that is by far good enough (and we don't need additional confusion; its already complicated enough).

Some more astronomy

At 21st of December we have on the northern hemisphere the shortest solar day and if somebody is situated on the southern tropic (that is 23.5° South) then he will see the sun at noon directly above himself. At 21st of June we have the longest solar day on the northern hemisphere and if somebody is situated on the northern tropic (that is 23.5° North) then he will see the sun at noon directly above himself.

So we come to the question where somebody in Hamburg will see the Sun the 21st of June at noon. For answering this question we imagine a triangle. The first corner is in the center of the Sun; the second corner is in the center of the Earth and the third corner we position on the surface of the Earth on the northern tropic. From our math lessons we remember that the sum of all angles in a triangle is 180°. Since our observer can see the sun directly above himself and the center of the Earth is directly below him, the angle at the observer is 180°.

Now we move the third corner away from the tropic in the direction of Hamburg (which is situated at 53° North). The distance to the Sun is very large (about 149 million kilometers) and the diameter of the Earth is relatively small (12,700 kilometers). Therefore the angle within the Sun will practically not change (from exactly zero to extremely near to zero). When we move the observer in direction of Hamburg then the angle there will get smaller and exactly by the same amount the angle at the center of the Earth gets bigger. In order to get the height over the horizon we have to subtract the difference of the angles from 90°. Alpha = 90° - (53° - 23,5°) = 60,5°. On the 21st of June at noon the Sun can be seen 60,5° over the horizon.

Now we know the height of the Sun over the horizon for two days of the year (better than nothing) but what about the other days? We could go the experimental path and measure for any day of the year the altitude where the sun is directly above oneself at noon. On the other hand we know that the length of the days changes slowly in winter and summer time while it changes rapidly in Spring and Autumn. The reason is that the latitude where the Sun can be seen directly above oneself changes with the sine of the angle where the Earth can be seen from the Sun.

We define that the angle of the Earth on the 21st of June as 0°; then in Autumn it is at 90°, in Winter at 180° and in Spring at 270°. Now we need to convert the days into degrees. The year got 365 days and a circle has got 360°; so one day equals 0,986° on the orbit of the Earth. The area where the sun could be seen directly above oneself is between +23,5° and -23,5°. Therefore the formula for calculation of the angle where the Sun will be seen at noon directly above an observer is (the nearly correct value is 23,4385°):

$$\text{Delta} = 23{,}5 * \sin(T * 360 / 365).$$

In this formula T is the number of days after the 21st of March because this is the day when the sun is directly above the equator at noon (and the sine of zero is zero). About 91 days later we got the 21st of June and the sine of 90° is 1. Therefore the Sun will be at 23,4385° latitude directly above an observer at noon.

Since somebody in Hamburg sees the Sun on the 21st of March at noon at an angle of 90° - 53° = 37°, we are now able to calculate the angle of the sun over the horizon at noon for any place in the world (for the southern hemisphere we have to replace the '+' by a '-'):

$$\text{Alpha} = (90° - \text{Latitude}°) + 23{,}4385 * \sin(T * 360 / 365)$$

The daily path of the Sun

First we consider the speed with which the Sun moves over the sky (well, this way of seeing at it was only valid until the Middle Ages; it is the Earth which rotates once per day around itself). The average distance between Earth and Sun is 149 million kilometers so the diameter of the Earth can be neglected. Physically correct (but not mathematically) we could let the world vanish and we ourselves float freely in space.

First the simple version of our situation. We have an ecliptic of zero degrees and we rotate once per day around the axis which goes from the front through our stomach to the back; and we tilt our head back so that we (from our point of view) look straight up. We have a panoramic 360° vision but we can't see anything below our feet.

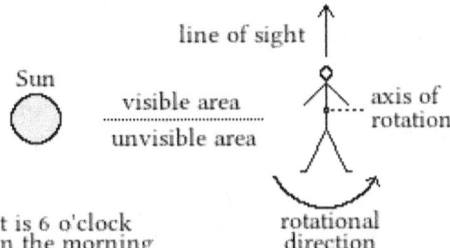

Now we define that our right side is East and our left side is West. Directly above us is South. Since we reduced the whole planet down to us we are naturally on the height of the equator (the center of the equatorial circle is within our stomach).

Since we can either see the sun or nothing at all this picture describes our situation perfectly. We can assume the size of this semicircle as we like; the sun could be as big as a pea and be just out of reach or it could be 149 million kilometers away and have its original size. We would not be able to tell which interpretation would be right. The numbers around the semicircle give the time of the day (18 being 6 o'clock p.m.).

Now we have a look at this semicircle from the side. To understand the sense of this picture we have to imagine that we have a solar panel in front of our feet. At 6 a.m. and at 6 p.m. 0% of the solar irradiation can fall on the panel. At 7 a.m. and at 5 p.m. one can see 25%. One hour later (earlier) it is 50%. So what we see in the picture is a graphical representation of the sine function.

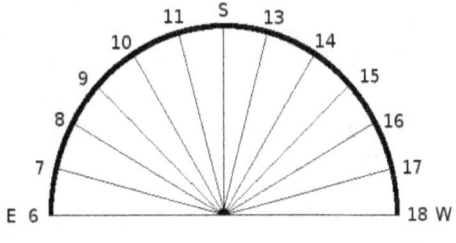

Now we introduce the next complication and that is our original position on the Earth. In the picture above we were laying on the equator disk with our stomach being the center. Now we have a look at how this picture changes when our stomach stays where it is but the orientation of our body is changed as if we were staying in Hamburg. If we raise our body by one degree then the rotational axis still goes through our stomach but we do a slight tumbling motion. The sun is not above us any more but moved by one degree.

If we simulate Hamburg then we have to turn the orientation of out body by another 52° (Hamburg is 53° North). Seen from the outside we do a strong tumbling motion but we are actually not interested what somebody else would see but what we would see. We are at the foot of the line which is tilted by 37° to the horizontal (90°- 53° = 37°). Our solar panel in front of our feet has done this motion as well and in the picture we see the result of the action. At the equator we had 100% yield at noon and now simulating Hamburg it is slightly more than 60%. Had we chosen Madrid (40° North, 3° West), then we would have about 76,6% of the maximum yield (again it depends on the sine of the angle).

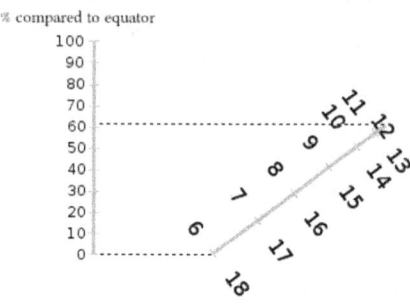

Now we come to the last important astronomical factor and that is the ecliptic. In the last example our body was tilted in respect of the rotational axis and we did some kind of tumbling movement. Now we leave our body in an angle of 90° to the axis but we tilt the rotational axis in respect to the ellipse of the path around the Sun.

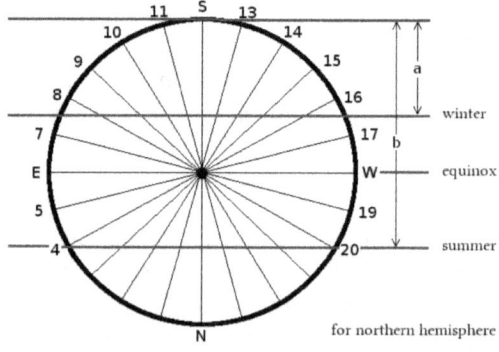

What we would now see in front of our eyes is in this picture. The tumbling movement is replaced for the ecliptic which has the effect that the sun seems to have a

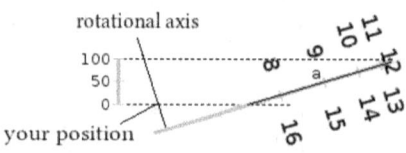

It is winter and the sun is about 20° over the horizon

298

different path any other day. When the solar day and night are equally long (equinox) then it looks as in the first example. Sun raise at 6 o'clock a.m. and sun down at 6 o'clock a.m. but during all other days the time differs. During winter time the solar days are much shorter and during summer time they are much longer.

We had already seen that the sun swings between the southern and the northern tropic. If again we have a sidewise look at the hour circle then we see that it is not only the angle which changes but the size of the visible sector of the Suns path as well (just to remind you: we are still looking straight up and we still have that solar panel in front of our feet.

In winter time the people in northern Germany see the Sun just slightly above the horizon. In the picture for the winter sun rise is at about 7:30 a.m. and the sun set is at about 4:30 p.m.. In the picture we have again the distance 'a' which we had seen in the hour circle. In the next picture we have a similar picture for a summer day. The sun will be much higher over the horizon and it can be seen during many hours more. Here again we find the distance 'b'.

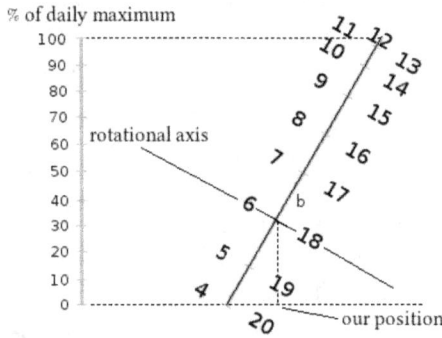

It is summer and the sun stays about 60° over the horizon

We already got a formula for calculating under which angle above the horizon the sun can be seen any single day at noon. Additionally we know that it turns 15° per hour. What is still missing is a formula to compute at which time the sun rise will be (since noon is exactly midday then we will know the time of sunset as well).

There is a relatively simple formula to compute (in hours) the time between sunrise and noon (I couldn't find the derivation of this formula; if you know it and if it is not too complicate, please let me know).

$$\text{Time}_{\text{Half\_day}} = \arccos(-\tan(\text{Longitude}°) * \tan(\text{Delta})) / 15$$

We just check this formula for Hamburg and equinox. Delta (the actual derivation between equator and and the place with the sun above oneself at noon in degrees) is zero and the tangent of zero is zero as well. The arc cosine of zero is 90 and 90 / 15 is 6. So the solar morning is 6 hours long and the solar afternoon is 6 hours long. That is, we have 12 hours of sunshine => equinox. During the winter Delta is -23.5° and we get the length for a solar day with 7.3 hours. In summer time we have Delta = +23.5 and a length of the solar day of 16.7 hours. The results and the personal experience fit well together.

Now we got a set of tools to calculate for any day during the year how high the Sun will appear over the horizon, we can calculate at what time sunrise and sundown will be and for any moment during the day we can find out in which direction we can find it at the sky. What we actually don't know is how high over the horizon the sun will be in the time in between.

For the derivation of the height angle (we only will need it for any full hour because to make calculations with an even finer resolution doesn't make much sense) we make a simplification and that is that we are not going to work with circle segments but with straight lines instead.

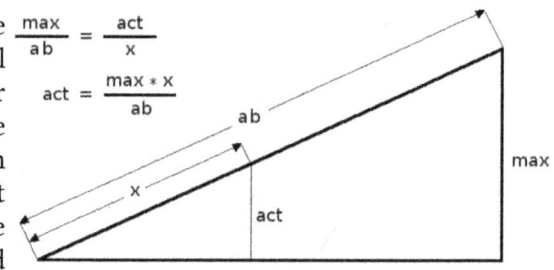

In order to get the actual elevation angle we are going to use the rule of three. 'max' is the elevation angle of the sun at noon. 'ab' is the distance we already know from the pictures; this value we take from the table coming soon. From the same table we will get 'x'. And with that we can know the actual elevation angle 'act'.

The distance 'ab' represents the circle of the Sun between sunrise and sundown. From the other formula we know for example that the sunrise is at 5 o'clock. So we go with the 5 as an index into the table and find in the last column 1.258 as 'ab'. If you want to know the elevation angle of the Sun at 9 o'clock so you go with 9 as an index into the table and you get from the last column 0.293. So x = 1,258 - 0,293 = 0,965. We got everything we need in order to get the elevation angle.

Time	Deviation South	abs(cos(x))	Delta	Distance 'ab'
12	0°	1.000	0.000	0.000
11 or 13	15°	0.965	0.035	0.035
10 or 14	30°	0.866	0.099	0.134
9 or 15	45°	0.707	0.159	0.293
8 or 16	60°	0.500	0.207	0.500
7 or 17	75°	0.258	0.242	0.742
6 or 18	90°	0.000	0.258	1.000
5 or 19	105°	0.258	0.258	1.258
4 or 20	120°	0.500	0.242	1.500
3 or 21	135°	0.707	0.207	1.707
2 or 22	150°	0.866	0.159	1.866
1 or 23	165°	0.965	0.099	1.965
0	180°	1.000	0.035	2.000

Global irradiation = direct irradiation + diffuse irradiation

In the section about the orbit of the Earth we had seen that in the years average 1,367 Watts can be received on one square meter of panel outside the atmosphere. This value was measured directly with satellites and should be pretty correct. If you wanted to measure the irradiation on the surface of the Earth then you might place a sensor at the end of an intransparent tube and aim for the sun.

The first thing you will observe is that the measured value is far below the 1,367 Watts. Additionally you will observe that you get quite a lot of irradiation from all parts of the sky and from all objects (if it were not like that you would see everything except the sky as pitch black). The amount of irradiation depends on the time of the day as well. How to explain this effect?

The reason is that the atmosphere is by far not as transparent as we perceive it. When a photon coming from the sun hits any kind of particle, an atom or a molecule, then first the energy is absorbed and normally emitted the next instant. With very small particles we can assume that the direction into which the energy is retransmitted depends on pure chance; so about 50% of that energy will again be on its way to the Earth. With bigger particles the chances are higher that the energy is retransmitted into the sky. The diffuse irradiation received from the sky near to the Sun will be higher than from those parts of the sky being near to the horizon opposite to the sun.

Since this diffusion happens everywhere within the atmosphere the light comes from all the visible sky. So we have the direct radiation (these are those photons which had no collision on their way) and the diffuse radiation (these are those photons which had one or more collisions). It is common to make the simplification that the diffuse irradiation from all parts of the sky has the same strength (in reality the diffuse irradiation is strongest near to the position of the sun and gets lower with the distance; but this effect is not strong enough to justify the additional effort. So the received irradiation is the sum of the direct and the diffuse irradiation.

When we have a look at the ratio between the two irradiations then we find out that this ratio depends on the location. In Hamburg the diffuse irradiation (over the year) is 57%, in Munich it is 54% and in Lisbon 35%. If we want to get valid data for the planing of a photovoltaic installation we have to find a way of how to take the diffuse irradiation into account.

How to get an estimate of the diffuse irradiation

A (seemingly) simple solution would be to set up another database for the diffuse irradiation. The problem is that worldwide there are relatively few places where the diffuse irradiation is measured with acceptable accuracy; the only way out is to derive these data from known data and that we can do by ourselves since there are some formulas which are suitable for that job and they base on the extraterrestrial irradiation.

First thing is to know how the extraterrestrial irradiation is defined. The global irradiation was the sum of all irradiations falling onto one square meter of flat ground. The extraterrestrial irradiation would be the irradiation falling onto this square meter if there were no atmosphere or if we lifted up this area (without changing the orientation) until it is outside the atmosphere (that's why extraterrestrial).

We know that on one square meter with optimal orientation 1,367 Watts are falling. What we have to do is to calculate how much of such a square meter could be seen by an observer looking from the sun. This, multiplied with 1,367 Watts, gives us the extraterrestrial irradiation. This correction factor changes over the day and is repeated only similar the next day (because the Earth followed its path by one degree).

The fact that extraterrestrial irradiation, global irradiation and diffuse irradiation are somehow interconnected is displayed in the next diagram (taken from the doctoral dissertation of Andreas Gassel, 1996).

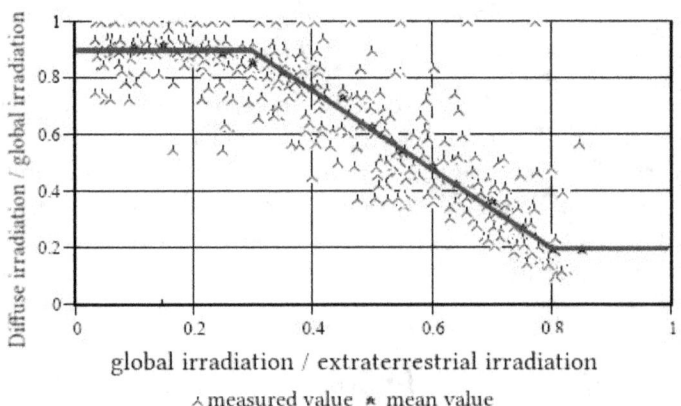

global irradiation / extraterrestrial irradiation

⋏ measured value ∗ mean value

In this dissertation a relatively complicated equation was derived from this diagram but we like it simple (if possible) so I defined a function with three sectors.

The definition of the function is as follows:

$P_{diff} = 0{,}9 * P_{glob}$ for $P_{glob} / P_{extra} < 0{,}3$

$P_{diff} = 0{,}2 * P_{glob}$ for $P_{glob} / P_{extra} > 0{,}8$

$P_{diff} = (-1{,}45 * P_{glob} / P_{extra} + 1{,}35) * P_{glob}$ for all other input values

Insufficient resolution of the irradiation data

The just developed function would be perfect to calculate for any hour of the day the ratio between direct and diffuse irradiation. The problem is that we only got the irradiation for a complete day in the months average.

We can calculate the extraterrestrial irradiation correctly for any moment of the day but we don't know the global irradiation, not even by the hour. The best what we can do in this situation is to calculate for the fifteenths of the actual month the extraterrestrial irradiation for any hour with sunlight (clouds neglected) and distribute the global irradiation with the same ratio over the day.

Our derived function for the ratio between global and diffuse irradiation can't deliver any differences over the day (because we only have the months average). So we have to assume that at least the tendency is more or less correct. We will not be able to correctly calculate the energy input for a North facing wall in Hamburg or use these data for an estimate in an economic feasibility study. But the result can give us an idea of what will happen.

The course of the calculation

We have a loop over all the months of the year.

For the 15th of the month we calculate the position of the sun for any full hour and the length of the solar day. Now we know for which hours we need to do the actual calculation (we only take the hours into account when the sun is shining 60 minutes).

Now we make a loop over the hours with sunshine.

First is now to calculate the extraterrestrial irradiation (that is the irradiation falling on one square meter in case that there is no atmosphere). This calculation we have to do for any hour since the value changes due to the rotation of the Earth.

Now we go into the data base and get the global irradiation per day (average over the month per square meter flat ground per day). Now it becomes a bit difficult since we only know the value for the whole day.

We can assume (with some qualification) that the global irradiation changes in lockstep with the extraterrestrial irradiation.

From the extraterrestrial irradiation we get the information how many percent of irradiation come down onto the ground during the actual hour. With this percentage we distribute the irradiation over the day (by doing so we compensate the error which we got by delimiting the calculation to those hours with sunshine).

The result is the global irradiation with its (probable) distribution over the day. With this information we use the derived function (the one with the three parts) and calculate the diffuse irradiation falling on one square meter of flat ground. Because of the simplification that the diffuse irradiation is evenly distributed over the sky we know, that a panel lying flat on the ground will get 100% of the diffuse irradiation (it can see all the sky) and a panel staying upright (even when facing North) will receive 50% of the diffuse irradiation because it can see exactly half the sky. So for any degree of inclination the panel will receive 0.55% less of the diffuse irradiation a flat lying panel would receive (the orientation of the panel has got no influence on this).

The last step will be to calculate for the different hours the direct irradiation. The direct irradiation for one square meter of flat ground is simply the difference between global and diffuse irradiation. Beware, this is only true for panels lying flat on the ground.

In order to calculate the irradiation on a panel with inclination we have to calculate the size of the shadow that the panel would produce. So we get as a factor how much bigger the shadow is compared to the value of a flat lying panel. If the sun is shining on the back of the panel then the percentage is zero. In case that the shadow is six square meters big then the percentage is 600% (no matter how big the panel actually is). Now we have to multiply the direct irradiation with this percentage and get the complete irradiation on the panel and this irradiation is going to vary strongly in the course of the day.

If you need more electric energy during winter time then you should install the panels as steep as possible so that their shades get as long as possible. Normally one will select the inclination so that one gets the maximum energy out of the panels; but there is the option to minimize the variations over the year.

Blank page for notes

www.ingramcontent.com/pod-product-compliance
Lightning Source LLC
Chambersburg PA
CBHW050049230526
45470CB00004B/1454